Digital Fourier Analysis: Advanced Techniques

Ken'iti Kido

Digital Fourier Analysis: Advanced Techniques

Springer

Ken'iti Kido
Yokohama-shi
Japan

Additional material to this book can be downloaded from http://extras.springer.com/

ISBN 978-1-4939-1126-4 ISBN 978-1-4939-1127-1 (eBook)
DOI 10.1007/978-1-4939-1127-1
Springer New York Heidelberg Dordrecht London

Library of Congress Control Number: 2014940156

Title of the Japanese edition: ディジタルフーリエ解析 2 基礎編—published by CORONA PUBLISHING CO., LTD—Copyright 2007

Printed on acid-free paper

Springer is part of Springer Science+Business Media (www.springer.com)

Preface

Fourier analysis is one method of investigating the origin of functions and their properties by using Fourier series and Fourier transform(s). The Fourier series and Fourier transforms were introduced by the mathematician Jean Baptiste Joseph Fourier at the beginning of the nineteenth century; they are widely applied in the engineering and science fields. One-dimensional waveforms (as functions of time or position) and two-dimensional images (as functions of two positional axes) are the main subjects studied nowadays using Fourier analysis. Fourier analysis is an important method used to analyze complex sound waveforms in the field of acoustical engineering. This is the reason why this book was originally listed in the series of books published by the Acoustical Society of Japan. However, since Fourier analysis is also applicable in many other engineering sciences, these two books, *Digital Fourier Analysis: Fundamentals, and Digital Fourier Analysis: Advanced Techniques*, are useful to readers in broader fields.

The Fourier transform itself does not fit well with analog processing because it requires too much numerical processing, such as multiplications and summations. That is the reason waveform analysis during the analog age could not make full use of the Fourier method of analysis. Fourier analysis was more valuable as the basis of theoretical analysis than for its practical applications during the analog era. However, the digitally processed Fourier transform became a reality with the emergence of the digital computer in the middle of the twentieth century. The later development of the Fast Fourier transform (FFT) algorithm in 1965, and the subsequent inventions of microchips for signal processing accelerated the application of Fourier analysis-based signal processing.

In the twenty-first century, Fourier analysis technology is widely used in our daily activities, and as a natural consequence, the technology is "hidden in a black box" in most of its applications. Even experts in the field use this technology without knowing details of how Fourier techniques are implemented. However, engineers, who wish to play important roles in developing future technologies, must do more than just deal with black boxes. Engineers must understand the basis of the present Fourier technology in order to create and build up new technologies based on it.

This book is intended so that high-school graduates or first or second grade college students with basic knowledge of mathematics can learn the Fourier analysis without too much difficulty. In order to do that, explanations of equations start from the very beginning and details of derivation steps are also described. This is very rare for this kind of specialized book.

This book also deals with advanced topics so that engineers presently involved in signal processing work can get hints to solve their own specific problems. Ways of thinking that lie behind or lead to theories are also described that are a must to apply theories to practical problems.

This work comprises two volumes. Seven chapters are included in Volume I, titled "Digital Fourier Analysis: Fundamentals." Volume II, titled "Digital Fourier Analysis: Advanced Techniques" contains six chapters. As the titles indicate, more advanced topics are included in Volume II. In this sense, the former may be classified as a text for undergraduate courses and the latter for graduate courses. Notice, however, that Volume I includes some advanced topics, whereas Volume II contains necessary items needed for a better understanding of Volume I.

Following is a brief explanation of each chapter. First, the contents of Volume I are briefly described.

Chapter 1 commences with an explanation of the impulse as being the limit of the summation of cosine waves with ascending frequencies. This chapter shows that all waveforms can be synthesized by use of Fourier series, i.e., that sine and cosine waves are the basis of waveform analysis. It then gives a geometric image to Euler's formula by explaining that the projections of a constantly rotating vector around the origin of the rectangular coordinate system onto their real and imaginary axes are the cosine and sine functions, i.e., the real and imaginary parts of a complex exponential function, respectively. The reader will be naturally guided to the concepts of instantaneous phases and instantaneous frequencies through this learning.

Chapter 2 starts with the theory on how to determine coefficients of the Fourier series of a periodic function. It investigates the properties of the Fourier series, showing why high order coefficients are needed for waveform synthesis and what kind of properties the Fourier series of even and odd function waveforms have. Then it shows that the Fourier series expansion becomes the Fourier transform pair when the period is made infinitely large.

Chapter 3 deals with problems encountered when one tries to express a continuous waveform by a sequence of numbers in order to numerically compute the Fourier transform. For that purpose, it investigates the most important issue in digitization: how to handle the sampling time based on the knowledge of the Fourier series; and guides the reader automatically to the sampling theorem. Then the relation between a continuous waveform and the discrete numerical sequence, which is the sampled version of the waveform, is discussed.

Chapter 4 guides you to the definition of discrete Fourier transform (DFT) and inverse Discrete Fourier transform (IDFT). The DFT transforms a numerical sequence with a finite length to another numerical sequence with the same length, and the IDFT transforms the latter sequence back to the former sequence. Then this chapter clarifies that these sequences are periodic and they have the length (data number) of the sequence as their periodicity. For later applications, the Discrete Cosine transform (DCT) is derived from basic concepts. The DCT describes the relationship between time domain and frequency domain functions using only cosine functions.

Chapter 5 explains a principle of the Fast Fourier transform (FFT) which drastically decreases the number of multiplications and summations required in the computation of the DFT. The FFT is an innovative numerical calculation method which has greatly expanded the range of application of Fourier analysis.

Chapter 6 discusses several items such as: (1) properties of the spectrum given by the DFT of a N-sample sequence (waveform) taken from a long chain of data; (2) its relation with the spectrum of the original data; (3) the relationship between sampling time and frequency resolution, and (4) the reason why new frequency components that do not exist in the original waveform are produced by the DFT, and so on.

Chapter 7 studies details of various weighting functions (time windows) applied to waveforms in order to obtain stable and accurate spectra (frequencies and amplitudes) by the DFT approach.

The explanation of Volume I ends here. But, as recognized by readers, the description is insufficient for use of Fourier Analysis in practical applications. More knowledge described in Volume II will be required for in-depth understanding of the descriptions in Volume I and for the application of Fourier analysis to a wide area.

The first chapter of Volume II guides the reader through the use of a convolution of two sequences to be calculated from an input and the system's impulse response. It becomes clear that the Fourier transform of a convolution can be expressed as a multiplication of the two respective Fourier transforms, and this leads to the exploration of a new way that a convolution in the time domain can be calculated in the frequency domain. Since an issue based on DFT periodicity is raised at this time, this chapter discusses the issue and explains in detail how to obtain the correct result.

In Chap. 2, the correlation function, that quantitatively expresses the degree of similarity between two time sequences, is derived. The difference between the correlation and convolution functions is that the directions of the time axes of one of two time sequences in the process of multiplication and summation calculation are opposite to each other. The Fourier transform of the correlation function of two time sequences is given as a cross-spectrum of the two functions in the frequency domain, which will be discussed in the next chapter.

Chapter 3 introduces a Cross-spectrum method that uses multiplication of spectra. The Cross-spectrum technique is a powerful method for uncovering an original function as an inverse process based on the convolution or the correlation function. This is a good example of the DFT's usefulness. DFT periodicity is the most important factor. This chapter illustrates the kind of problems related to Cross-spectrum analysis, and discusses how to avoid errors, by taking advantage of periodicity.

Chapter 4 introduces the concept of a Cepstrum which is defined as a Fourier transform of the logarithm of a spectrum. This useful method of analysis is based on a little quirky idea. Cepstrum analysis is a powerful method of signal analysis for detecting hidden information that is not visible from the Fourier transform of a time domain signal.

Chapter 5, at first, analyzes the problem that occurs when a waveform is depicted as a rotating vector in order to obtain its envelope. Since an orthogonal waveform of the original waveform is needed to get a rotating vector, the question of how to derive the orthogonal function in the frequency region is discussed. The calculation of the orthogonal function in the time region by applying the inverse Fourier transform to the orthogonal function in the frequency region results in the Hilbert transform. While the Hilbert transform is a demodulation of the amplitude-modulated wave to get an envelope as a length of rotating vector, it is also, a demodulation of the frequency-modulated wave, producing an instantaneous frequency which is the rotating speed of the vector.

Chapter 6 touches upon two-dimensional DFT and DCT methods. At first, a definition of the two-dimensional DFT is given. When the reader tries to obtain two-dimensional spectra of images with basic patterns, one can easily guess its output from the relation between the one-dimensional waveform and its spectrum. The reader will understand that one-dimensional Fourier transform procedure is an important base. By showing definitions of DCT and samples of two-dimensional DCT spectra of simple images, the concept of how information compression by DCT takes place will be explained with concrete examples.

One of the features of this book is that it contains a number of figures that have an interactive supplement. (The supplementary files can be downloaded from http://extras.springer.com). Figures with this feature are indicated by their caption, which includes the file name of the corresponding animation file. To view the animation, click the corresponding exe file to start the program, and then click the green "start" button after data input and/or selection of conditions. Then the program starts the calculation based on the theory, input data, and conditions. The reader may see unexpected results occasionally. As they have their own causes or reasons, it will be worthwhile for the reader to think of them for a deeper understanding. Note that the programs are written in Visual Basic and may not work on all computers.

I would like to emphasize the following through my long experience as a writer of this book and also as a user of this book in my classes and other lectures. The reader will lose more than he/she earns if he/she prematurely thinks that he/she has understood one topic after running a related program and briefly looking at the result. The reader must run programs with various data and conditions and look at the corresponding results and then he/she must think how they are related with each other. With the attached programs, the reader can do these easily while having some fun.

Very few references are listed at the end of this book compared to the contents of this book. This is because most of the theories are described from the beginning, and as a result this book became self-contained. Theory-oriented readers should refer to books such as [4–9] in Reference. Since the Fourier analysis techniques are developing day by day, the readers should refer to current journals in the related area.

Finally, although I would like to express my sincere appreciation to all of those who gave me tremendous encouragement and cooperation to write this book, I must apologize that I cannot list up all of their names. My excuse is that so many people assisted me in writing this book.

This book, originally published in Japanese, was translated first by Dr. Hideo Suzuki, a former professor of Chiba Institute of Technology, Mr. Jin Yamagishi, a technology management consultant of JINY Consultant Inc., and myself, and then, very carefully checked and corrected by Dr. Harold A. Evensen of Michigan Technological University in USA and Dr. Leonard L. Koss of Monash University in Australia.

I would like to express my sincere appreciation to all of those who contributed to publishing this book.

February 2013 Ken'iti Kido

Contents

Chapter 1
Convolution

If we know a system's *impulse response*, we can calculate the output of that system for any input. The impulse response is defined here as the output response to the *unit impulse* input to a linear system. The calculation of the output based on the input and the impulse response is called *convolution* defined in the time domain. The same calculation can be carried out in the frequency domain using the Fourier and inverse Fourier transforms. The reason we use this roundabout method is because the output is obtained in much shorter time than by directly calculating the convolution integral in the time domain.

In this chapter, the basis of the calculation to obtain the output waveform from the input waveform and the impulse response are explained first, then formulae to calculate the convolution are derived. It is also shown that the convolution of two functions in the time domain can be expressed as a multiplication of their respective Fourier transforms in the frequency domain. A digital *finite impulse response* (*FIR*) *filter*, one that produces a digital output of a system by convolution of the input and the impulse response of the system, is introduced as well as a determination of the filter coefficients and an output calculation method. At the end of this chapter, an efficient method of calculation in the frequency domain is given.

1.1 Convolution

Let's consider an "output calculation method" when an arbitrary input is applied to a transmission system with a known impulse response. A transmission system has an input port and an output port (or terminal) as shown in Fig. 1.1. The output response of the system when a unit impulse is given to the input terminal is the impulse response function. The purpose of this chapter is to derive a method to obtain an output when an arbitrary input sequence as shown in Fig. 1.1 is applied to the system.

The output of the system to a single pulse is equal to the impulse response multiplied by the amplitude of the single pulse. Since a sample sequence is considered for the calculation, the timing of the pulse input(s) corresponds to the

K. Kido, *Digital Fourier Analysis: Advanced Techniques*,
DOI: 10.1007/978-1-4939-1127-1_1,

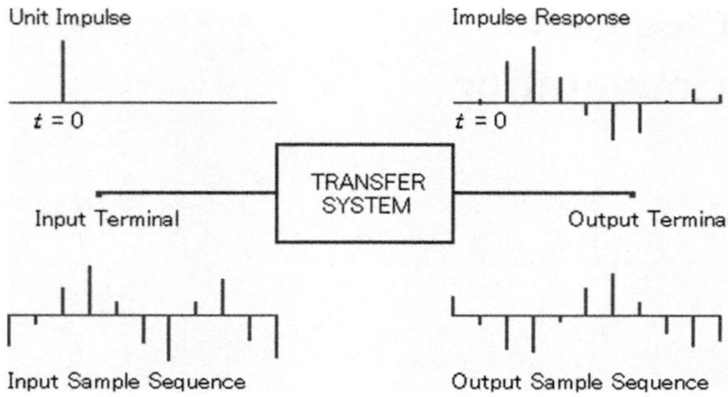

Fig. 1.1 Digital transfer system and its input and output signals

timing of the sampling and, therefore, the output sequence exists only at sampling times after the commencement of the input.

In realizable transmission systems, there is no output before the commencement of the input to the system. Systems that satisfy this property (causality) are called *causal systems*. Impulse responses of causal systems are always zero for $t < 0$.

The output response sequence of a system to the application of the first impulse has the same shape as the impulse response function, and the magnitude of the response sequence is proportional to the amplitude of the first impulse. The sequence begins immediately upon receiving the first impulse. The magnitude of the response sequence corresponding to the second impulse given at the next sampling time is proportional to the amplitude of the second impulse and delayed by one sampling interval after the first response sequence. These two response sequences are superimposed. The response sequences to the third and later impulses behave in the same way.

At each sampling time point, the responses to successive input pulses are superimposed if the system is linear. Each of the superimposed responses is given as the sum of the response sequences for each of the sampling time points.

Let's derive a formula to obtain the output impulse sequence, referring to Fig. 1.2, that demonstrates the process explained above. In the figure, $h(n)$ is the impulse response, and $x(n)$ is the input sample sequence. The horizontal axis is the discrete sample time axis represented by n.

First, the pulse $x(-3)$ is input to the system. The response to this input is $r(-3, n)$, where n starts from -3. The shape of $r(-3, n)$ is the same as the impulse response and the magnitude is proportional to the amplitude of $x(-3)$. Next, when the pulse $x(-2)$ is input, $r(-2, n)$ is the system response to it. The response to the pulse $x(-1)$ is slightly different. Its response shape $r(-1, n)$ is reversed because $x(-1)$ is negative. The next three inputs are positive, but since $x(3)$ and $x(4)$ are negative, response shapes to them are also reversed. The responses to the successive inputs are shown from the third row and below in the figure, with each of them starting at the time of the corresponding input. The amplitude of each response is proportional to each input pulse and the shape is proportional to the impulse response.

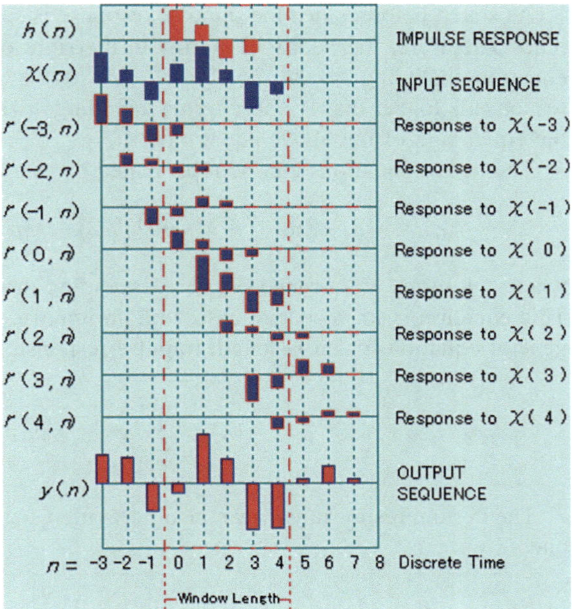

Fig. 1.2 Calculation of output $y(n)$ $(-3 \leq n \leq 7)$ by adding responses $r(m, n)$ $(-3 \leq m, n \leq 7)$ of the transfer system with the impulse response $h(n)$ $(0 \leq n \leq 3)$ to the input $x(n)$ $(-3 \leq n \leq 4)$. Animation available in supplementary files under filename E8-02_ConvolutionStepByStep.exe

Thus far, the responses to individual inputs have been shown. The output of the system is simply the summation of all responses, which is shown as $y(n)$ at the bottom of the figure.

The calculation of $y(n)$ demonstrated in Fig. 1.2 must be represented by equations. Since there are too many $y(n)$ to write them all down, let's express the first four—from $y(0)$ to $y(3)$.

At $n = 0$, the time of $y(0)$, there are four responses, the third pulse of the response, $r(-3, 3)$, to the input $x(-3)$, the second pulse of the response, $r(-2, 2)$, to the input $x(-2)$, the first pulse of the response, $r(-1, 1)$, to the input $x(-1)$, and the 0-th pulse of the response, $r(-0, 0)$, to the input $x(0)$. The response of the system $y(0)$ is the summation of these four terms. From the top, $r(-3, 3)$ is equal to $x(-3)h(3)$, $r(-2, 2)$ to $x(-2)h(2)$, $r(-1, 1)$ to $x(-1)h(1)$, and $r(0, 0)$ to $x(0)h(0)$. Therefore, $y(0)$ is given by

$$y(0) = x(-3)h(3) + x(-2)h(2) + x(-1)h(1) + x(0)h(0)$$

Equations for $y(1)$ and latter ones are given in the same way.

$$\begin{aligned}
y(1) &= x(-2)h(3) + x(-1)h(2) + x(0)h(1) + x(1)h(0) \\
y(2) &= x(-1)h(3) + x(0)h(2) + x(1)h(1) + x(2)h(0) \\
y(3) &= x(0)h(3) + x(1)h(2) + x(2)h(1) + x(3)h(0) \\
y(4) &= x(1)h(3) + x(2)h(2) + x(3)h(1) + x(4)h(0)
\end{aligned}$$

As stated before, since the impulse response is causal, $h(n) = 0$ for $n < 0$ must hold. Therefore, the additional terms to the right of the right-hand side of each equation including $h(-1)$, $h(-2)$, . . . do not exist. However, if the impulse response is longer than 4, more terms including $h(4)$, $h(5)$, . . . must be added past the equal sign of the above equations.

The above equations can be represented by one equation using $y(n)$ as

$$y(n) = x(n-3)h(3) + x(n-2)h(2) + x(n-1)h(1) + x(n)h(0)$$

where the terms are arranged in the descending order of $h(k)$ (i.e., $h(3) \rightarrow h(0)$). This equation is for a special case with an impulse response length of 4. A more general equation for an m-length impulse response is given by

$$y(n) = \sum_{p=0}^{m-1} x(n-p)h(p) \tag{1.1}$$

The system response $y(n)$ can also be written so that k in $x(k)$ increases one by one.

$$\begin{aligned} y(n) = \ldots &+ x(-2)h(n+2) + x(-1)h(n+1) + x(0)h(n) + x(1)h(n-1) \\ &+ x(2)h(n-2) + x(3)h(n-3) + x(4)h(n-4) + x(5)h(n-5) + \ldots \end{aligned}$$

Since the impulse response $h(n)$ is limited in $0 \leq n \leq m-1$ (response length m), the above equation is rewritten as:

$$y(n) = \sum_{p=n-m+1}^{n} x(p)h(n-p) \tag{1.2}$$

Equations (1.1) and (1.2) are both convolutions of $x(n)$ and $h(n)$. These two equations appear different from each other. But, since both are derived using Fig. 1.2, they represent the identical operation. In fact, one of them can be derived from the other, as shown in Appendix 8 (A).

Since it is laborious to use Σ each time to represent the convolution operation, it is a common practice to represent the convolution in the form of Eq. (1.3).

$$y(n) = x(n) * h(n) \quad == h(n) * x(n) \tag{1.3}$$

Thus far, in order to make the story clearer, and to relate directly to the numerical calculation, the input and output signals have been treated as digital signals. However, for future use, a representation for continuous systems will also be given here.

There is no change in Fig. 1.1 even if the system is changed from a discrete system to a continuous system. However, the input, output, and the impulse response are continuous functions. Therefore, the discrete multiplications and summations in Eqs. (1.1) and (1.2) are represented by the integrals of continuous functions with the impulse response length T and an infinitesimally small sampling interval. Equations (1.1) and (1.2) in the discrete system are now replaced by Eqs. (1.4) and (1.5) in the continuous system, respectively.

$$y(t) = \int_0^T x(t-\tau)h(\tau)d\tau \tag{1.4}$$

$$y(t) = \int_{t-T}^{t} x(t)h(t-\tau)d\tau \tag{1.5}$$

In Eqs. (1.4) and (1.5), $h(t)$ is given physical meaning as an impulse response function and, therefore, the integration ranges are restricted within $0 \sim T$ and $t-T \sim t$, respectively. However, $h(t)$ can be any function, and the parameter t does not have to be time. Considering it in this way, the limitation of causality can be removed and the more general formulae for convolution are given in Eqs. (1.6) and (1.7).

$$y(t) = \int_{-\infty}^{+\infty} x(t-\tau)h(\tau)d\tau \tag{1.6}$$

$$y(t) = \int_{-\infty}^{+\infty} x(t)h(t-\tau)d\tau \tag{1.7}$$

The same symbol * is used for the convolution in the continuous system as well as in the discrete system.

$$y(t) = x(t) * h(t) \tag{1.8}$$

1.2 FIR Digital Filter

If the impulse response of a system is known, the system's output response to a given input can be calculated by convolution. For a continuous system, this calculation is sometimes very difficult. However, if the impulse response is a sample sequence, the convolution between the input sample sequence and impulse response sequence can be calculated using Equations (1.1) or (1.2). In this case, the sequence representing the impulse response must be finite. This is almost always satisfied since real impulse responses gradually decrease as time passes after the end of the input sequence. In these cases, the impulse response lengths can be made finite by truncating the responses after they become negligibly small.

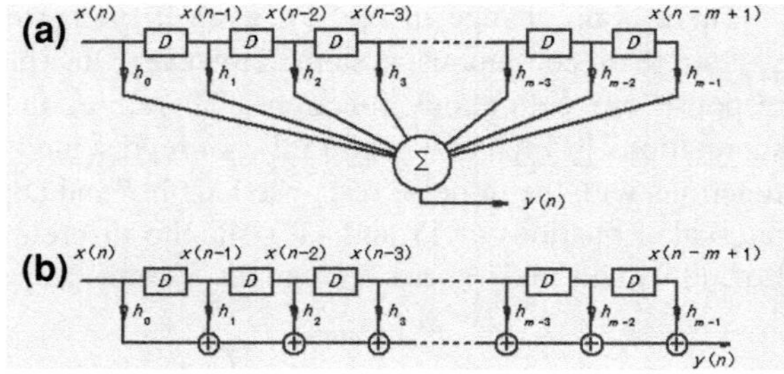

Fig. 1.3 Flow of signals in the FIR digital filter

A digital filter or a system that performs a convolution between the finite impulse response and the input sequence is called an *FIR digital filter*.

The flow of signals in a digital filter can be represented by either Fig. 1.3a or b. In Fig. 1.3a, the summation is accomplished at one time, and in Fig. 1.3b, the summation is done successively, but actually they accomplish the same result. The input is $x(n)$ shown at the top left, and n increases by one with every sample time. The symbol D denotes a *delay unit*. The output of the left-most delay unit is $x(n-1)$, and that of the second left-most unit is $x(n-2)$, and so on. After going through $(m-1)$ delay units, the output is the sample value sampled at the $(m-1)$ sample time.

The arrow indicates the multiplication of the delayed signal coming from the delay unit by the coefficient shown at the right of the arrow $(i = 0 \sim m-1)$ where m is the number of the delay units.

Below the arrow, the two charts are different. The symbol $\oplus$ indicates the process of summing the two signals from the left and the top and giving output to the right. The time needed for a signal to go through one delay unit is one sample time. If the multiplications and summations are accomplished much faster than the one sampling time, the output can be renewed at every sample time.

Figure 1.3b, which successively adds two data, may seem more practical than 1.3a, which adds many data at one time. However, in real processing, many multipliers and adders are not used, but it is common practice that inputs to a fast signal processor are changed successively and the necessary calculations are accomplished.

From the above discussion and Fig. 1.3, it can be seen that the output is given by

$$y(n) = x(n)h_0 + x(n-1)h_1 + x(n-2)h_2 + \cdots + x(n-m+1)h_{m-1} \tag{1.9}$$

Since this is exactly the same as Eq. (1.1), the process described in Fig. 1.3 is the convolution between $x(n)$ and $h(n)$. The operation in Fig. 1.3 is considered a process of averaging $x(n)$ over the past m values with weighting factor $h(n)$ and, therefore, it is called a *moving average (MA) filter*.

The important problem is how to determine the coefficients $h(n)$ in order to give a desired frequency response (characteristic). One solution is to obtain the impulse response following the established method of the analog filter circuit design, and to use its sample values. However, this process requires many tedious steps. By using digital signal processing, without depending on the conventional methods, a digital filter with arbitrary frequency response can be easily obtained. One method will be shown later in this chapter.

Another important practical problem is to obtain the impulse response of a system, given the input to and the output from the system. This is a topic of Chap. 3.

1.3 Calculation of the Filter Output

As Eq. (1.9) shows, $(m + 1)$ multiplications and summations are necessary to obtain one sample value. This calculation must be accomplished in one sample time.

In the real world, the frequency ranges of sound we hear can be over the whole audio range and the impulse response can have a duration exceeding 1 s. For example, the reverberation times (time needed to decay to 1/1,000 from the maximum amplitude) of concert halls and vibrations of musical instruments and metallic structures sometimes reach several seconds. In order to calculate the voice, musical, or other acoustical/vibratory signals via those systems, tens or hundreds of thousands of multiplications and summations are necessary to obtain one sample output. Digital signal processors (DSPs) have been developed for accomplishing fast and efficient multiplications and summations of signals. There is also a method of parallel processing that uses plural processors. Not only one-dimensional signals with sampling frequencies of tens of thousands of Hertz, but also signals with MHz sampling frequencies, and two- and three-dimensional signals. For these wide applications, there is no limit to the requirements for processing speed.

Examples of the convolution calculation are shown in Fig. 1.4. An impulse response of a system is shown in Fig. 1.4a, and two inputs are shown in (b) and (d). Figure 1.4a is an impulse response of a single degree-of-freedom system [see Appendix 8(B)]. Figure 1.4b shows a one-cycle sine wave at 0.4 times the system resonance frequency. Figure 1.4d shows a four cycle sine wave at 2 times the system resonance frequency. The system responses to these two signals are shown in Fig. 1.4c, e respectively. As the input frequency gets closer to (or farther from) the resonance frequency, the output amplitude increases (or decreases).

The program that generated Fig. 1.4 enables the user to check results for sine, cosine, rectangular, saw-tooth wave inputs with arbitrary lengths of sample sequences and sample frequencies.

In the above examples, n and p are in the range less than 100 and the size of the calculation is very small. With the impulse response of 32 samples and the input length of 16 samples, the number of multiplications and summations is 512. This can be performed in an instant by a modern personal computer. If the length of the impulse response becomes several thousands or tens of thousands, a method of reducing the calculation steps must be devised.

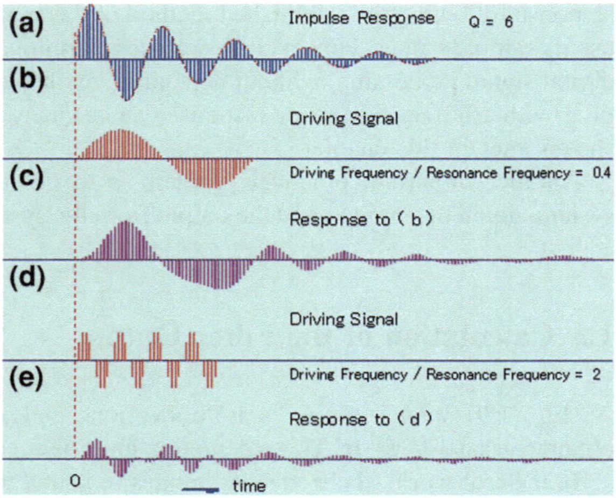

Fig. 1.4 Inputs and outputs of a transfer system. **a** Impulse response, **b** driving signal (frequency ratio: 0.4), **c** response to driving signal (**b**), **d** driving signal (frequency ratio: 2), **e** response to driving signal (**d**). Animation available in supplementary files under filename E8-04_VibInsulator.exe

1.4 Fourier Transform of the Convolution

Applying the FFT and accomplishing the main part of calculation in the frequency domain is a powerful method for reducing the computation load. Studying the FFT of the convolution integral is also beneficial for the understanding of convolution. First, consider the Fourier transform of the convolution of continuous functions.

Performing the Fourier transform of the convolution of $x(t)$ and $h(t)$ given by Eq. (1.6):

$$\mathrm{FT}[x(t) * h(t)] = \int_{-\infty}^{+\infty} \int_{-\infty}^{\infty} x(t-\tau)h(\tau)d\tau \exp(-j2\pi ft)dt$$

Using a new parameter $u = t - \tau$ $(t = u + \tau)$ and, if Fourier transforms of $x(t)$ and $h(t)$ exist, the above equation can be rewritten as:

$$\begin{aligned}
&= \int_{-\infty}^{+\infty} \int_{-\infty}^{\infty} x(u)h(\tau)d\tau \exp\{-j2\pi f(u+\tau)\}dt \\
&= \int_{-\infty}^{+\infty} \int_{-\infty}^{\infty} x(u)\exp(-j2\pi fu)h(\tau)\exp\{-j2\pi f\tau)dud\tau \\
&= \int_{-\infty}^{+\infty} x(u)\exp(-j2\pi fu)du \int_{-\infty}^{\infty} h(\tau)\exp\{-j2\pi f\tau)d\tau
\end{aligned}$$

Further it can be rewritten as:

$$\mathrm{FT}[x(t) * h(t)] = \int_{-\infty}^{+\infty} [x(t) * h(t)] \exp(-j2\pi ft)dt = X(f)H(f) \tag{1.10}$$

where the Fourier transform of $h(t)$, $H(f)$, is referred to as the *transfer function (or frequency response function)* of the system. Applying the inverse Fourier transform to both sides of Eq. (1.10), we have the equation below.

$$x(t) * h(t) = \int_{-\infty}^{+\infty} [X(f)H(f)] \exp(j2\pi ft)dt = \mathrm{IFT}[X(f)H(f)] \tag{1.11}$$

The above equation shows that the convolution of two functions in the time domain is accomplished by performing the inverse Fourier transform of the product of the Fourier transforms of the two time functions.

Considering the symmetry of the Fourier transform, the following equation can also be obtained.

$$x(t)h(t) = \int_{-\infty}^{+\infty} [X(f) * H(f)] \exp(j2\pi ft)df = \mathrm{IFT}[X(f) * H(f)] \tag{1.12}$$

The derivation of Eq. (1.12) is shown in Appendix 8(C). Of course, the relationship given by Eq. (1.13) also exists.

$$\mathrm{FT}[x(t)h(t)] = \int_{-\infty}^{+\infty} x(t)h(t) \exp(-j2\pi ft)dt = X(f) * H(f) \tag{1.13}$$

The above equations are valid for continuous functions, and they are also valid for discrete sample sequences.

First of all, an N-point DFT version of Eq. (1.1) is obtained. The impulse response length m must be shorter than or equal to N. If it is shorter, it is made equal to N by appending zero data to $h(p)$ $(m \le p \le N-1)$. Then, the discrete version of Eq. (1.1) is given by

$$y(n) = \sum_{p=0}^{N-1} x(n-p)h(p) \tag{1.14}$$

Calculating the DFT of Eq. (1.14):

$$Y(k) = \sum_{n=0}^{N-1} \sum_{p=0}^{N-1} x(n-p)h(p) \exp(-jkn/N) \tag{1.15}$$

A new parameter q is introduced here.

$$q = n - p \quad \text{or} \quad n = q + p$$

The ranges of p and n are from 0 to $N-1$. And since $x(n-p)$ is a periodic sequence with period N, the range of q is also from 0 to $N-1$. Then, the above equation can be rewritten as:

$$
\begin{aligned}
Y(k) &= \sum_{q=0}^{N-1}\sum_{p=0}^{N-1} x(q)h(p)\exp\{-jk(q+p)/N\} \\
&= \sum_{q=0}^{N-1} x(q)\exp(-jkq/N)\sum_{p=0}^{N-1} h(p)\exp(-jkp/N\}
\end{aligned}
$$

Using the symbol for the convolution operation, the above equation can be represented by

$$
Y(k) = \mathrm{DFT}[x(n) * h(n)] = X(k)H(k) \tag{1.16}
$$

It has been shown that $Y(k)$, the DFT of $y(n)$, which is the convolution of $x(n)$ and $h(n)$, is given by the product of $X(k)$ and $H(k)$, which are the DFTs of $x(n)$ and $h(n)$, respectively. Of course, the IDFT of $Y(k)$ is $y(n)$.

$$
y(n) = x(n) * h(n) = \mathrm{IDFT}[X(k)H(k)] \tag{1.17}
$$

An equation for the discrete-system equivalent to Eq. (1.12) for the continuous system can also be derived:

$$
\begin{aligned}
x(n)h(n) &= \frac{1}{N^2}\sum_{k=0}^{N-1} X(k)\exp(j2\pi kn/N)\sum_{p=0}^{N-1} H(p)\exp(j2\pi pn/N) \\
&= \frac{1}{N^2}\sum_{k=0}^{N-1}\sum_{p=0}^{N-1} X(k)H(p)\exp\{j2\pi(k+p)n/N\}
\end{aligned}
$$

Using a new parameter $r = k + p$ ($p = r - k$), the above equation is rewritten as:

$$
\begin{aligned}
x(n)h(n) &= \frac{1}{N^2}\sum_{r=0}^{N-1}\sum_{k=0}^{N-1} X(k)H(r-k)\exp\{j2\pi rn/N\} \\
&= \frac{1}{N^2}\sum_{r=0}^{N-1} [X(r) * H(r)]\exp\{j2\pi rn/N\} \\
&= \frac{1}{N^2}\mathrm{IDFT}[X(k) * H(k)]
\end{aligned} \tag{1.18}
$$

Now, the reader may question why the frequency domain is used to calculate the convolution integral, using the Fourier transform, which can be calculated directly in the time domain. The answer to this question is given by the fast Fourier transform (FFT) introduced in Chap. 5. The large number of multiplication and summation operations needed for the convolution calculation in the time domain is significantly reduced by use of the FFT operation in the frequency domain. The degree of reduction can now be discussed numerically.

First, check the case of directly calculating the convolution, following Eq. (1.1). In order to compare with the case using an N-point FFT, let the lengths of both $x(n)$ and $h(n)$ be N. Then the range of n in $y(n)$ is from 0 to $2N - 1$. Since we need N multiplications and summations to obtain one value of $y(n)$, it seems that the total number of multiplication would be $2N^2$. But this is not correct. The circles (○) in Fig. 1.5 show the cases when multiplication is needed for $N = 4$. For $0 \le n \le N - 1$ (=3) and for $N \le n \le 2N - 2$ (=6), $(n + 1)$ and $(2N - 1 - n)$ multiplications are necessary to calculate each $y(n)$, respectively. Therefore, the total number of multiplication to obtain $(2N - 1)$ values of $y(n)$ becomes N^2. The number of summations is less than that of multiplication, which is equal to $(N^2 - 2N + 1)$. The $(2N - 1)$ double circles (◎) in the bottom row indicate the addition of the products at each sample time, which gives the convolution of $x(n)$ and $h(n)$.

Next, check the number of multiplications and summations needed for the Fourier Transform method using Eqs. (1.16) and (1.17). For the FFT of each of $x(n)$ and $h(n)$, $(N/2)\log_2 N$ multiplications and summations are necessary. After N multiplication of the twos, it seems that an N-point IFFT is necessary to obtain $y(n)$ from $Y(k)$. However, the number of sample values obtained in this way is only N, whereas, the number obtained by the time-domain method is $2N - 1$. The reason is that the samples exceeding N, indicated by ⊙, are included in the area shown by △ because of the *circularity* (see Sect. 1.5) of the FFT. In order to avoid this problem, N zero data must be added to both $x(n)$ and $h(n)$ and their lengths must be made $2N$. Then, by changing N to $2N$, the total number of multiplications, $P(N)$, is given by

$$\begin{aligned} P(N) &= 2N\log_2 2N + 2N + N\log_2 2N \\ &= 3N\log_2 N + 5N \end{aligned}$$

When the convolution is calculated for one constant impulse response function and a long input sequence, one $2N$-point FFT is necessary for the impulse response. If the length of the input sequence is much longer than $2N$, the above equation becomes (the derivation is omitted):

$$P_R(N) = 2N\log_2 N + 4N$$

Table 1.1 shows a comparison of the number of multiplications needed for the direct calculation N^2 and the number given by the above equation for various values of N. The table gives the results up to $n = 1{,}024$. Note that the reduction ratio increases significantly as N gets larger.

The number of summations is also significantly reduced by use of the FFT method. The effect of reduction of operations is obvious for the data sizes given in Table 1.1. As the impulse response gets longer, the reduction ratio gets quite large.

For the FFT method, the N-point data set must be ready. But if you wait until it is ready, there is a delay of N samples even if the computation time is negligible. For applications such as adding reverberation to a live performance, this method is not suitable. However, if the time delay is permissible, the FFT method is very effective in reducing the calculation time.

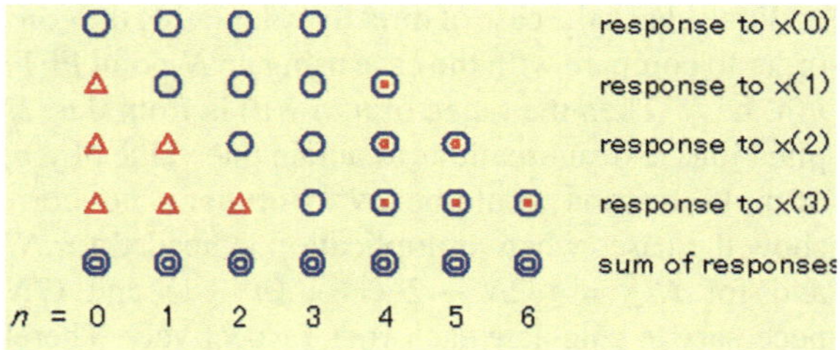

Fig. 1.5 Multiplication and summation operations in the convolution calculation of two N-point sequences (case for $N = 4$)

Table 1.1 Comparison of the numbers of multiplication operations necessary for the convolution calculation in the time and frequency domain

Data number N	Numbers of multiplication operations		
	Time domain calculation (N^2)	Frequency domain calculation	
		$3N \log_2 N + 5N$	$2N \log_2 N + 4N$ (data length $= \infty$)
8	64	112	80
16	256	272	192
32	1,024	640	448
64	4,096	1,472	1,224
128	16,384	3,328	2,304
256	65,536	7,424	5,120
512	262,144	16,385	11,254
1024	1,048,576	36,840	24,576

Other than the computation time reduction, Eq. (1.17) has an important meaning. It becomes possible to add a frequency dependent weight to the output. Some frequency ranges can be enhanced and other ranges may be suppressed. The weighting in the frequency range is very flexible in the FFT method. Studying Eq. (1.17), one may think that the input $X(k)$ can go through any arbitrary transfer function $H(k)$ by accomplishing the IFFT of $X(k)H(k)$. However, it must be remembered that the DFT/IDFT assumes periodicities both in the time and frequency domains. Applying the DFT/IDFT without keeping this fact in mind may lead to a serious problem. This concept will be discussed later in this chapter.

1.5 Circular (Periodic) Convolution

Consider Fig. 1.5 again. This figure shows the calculation of convolution by the FFT method for the case with $N = 4$. The responses for individual $x(n)$ denoted by $\odot$ ($n > 3$) are circulated back to the places shown by $\triangle$. As stated before, this is due to the periodicity of the DFT operation.

Figure 1.6 shows the situation in more detail. In this figure, two 16-point sequences are shown by (a) and (b) on the left, and their DFTs are shown by (pa) and (pb) on the right. The product of the two DFTs are shown by (pc) on the right and its IDFT is shown by (c) on the left. This has only 16 points. Since the convolution calculated by the direct method in the time domain has 32-point data as shown by (d) at the bottom, it is clear that (c) is not the correct convolution of (a) and (b).

Figure 1.6c is equal to the results of (d) from $n = 0$ to 15 added to the results of (d) from $n = 16$ to 31 (indicated by the white bars). The second N-point data that are pushed out from the first N-point data are overlaid with the latter because of the periodicity of the DFT.

As Fig. 1.6 shows, the convolution calculated by the N-point FFT method also has the periodicity with respect to N. Periodic data can also be displayed on a circle on the complex plane and in this case the periodic property becomes the circular property. Therefore, the convolution obtained by the FFT method is called *circular convolution.*

Since the convolution of two N-point sequences given by the N-point FFT has the length N, a correct answer with $2N$ points cannot be obtained. As stated in the discussion of number of operations in the FFT in Chap. 5, a $2N$-point FFT must be used in order to get a $2N$ point result. As was done in Sect. 1.4, N-point zeros are added to the N-point sequences of $x(n)$ and $h(n)$ and their convolution is calculated.

The five steps of the calculation are given below:

1. N-point zeros are added to both N-point sequences, $x(n)$ and $h(n)$.
2. $X(k)$, the $2N$-point FFT of $x(n)$ is calculated.
3. $H(k)$, the $2N$-point FFT of $h(n)$ is calculated.
4. $Y(k) = X(k)H(k)$ is calculated.
5. $y(n)$, the $2N$-point IFFT of $Y(k)$, is calculated.

Figure 1.7 shows the process of the above calculation and the results. Figure 1.7a, b are the same $x(n)$ and $h(n)$ as those of Fig. 1.6a, b. However, in Fig. 1.7, N-point zeros are added to them. Figure 1.7 (pa) and (pb) are $X(k)$ and $H(k)$, the $2N$-point FFTs of $x(n)$ and $h(n)$, respectively. Since the frequencies are not integers, the DFTs exhibit wide spreading. Even if the frequencies are integers, since the latter halves of the two sequences are zeros, the $2N$-point DFTs will always display spreading. The product (pc) of (pa) and (pb) includes mostly nonzero values. As expected, Fig. 1.7c given by the IDFT of (pc) is equal to (d), which is the convolution directly calculated in the time domain.

It has been made clear that the error due to circular convolution can be avoided, and convolution can be obtained by DFT/IDFT in the frequency domain. Of course, DFT/IDFT should be FFT/IFFT to reduce the computation time. The reduction of the number of multiplications is already given in Table 1.1.

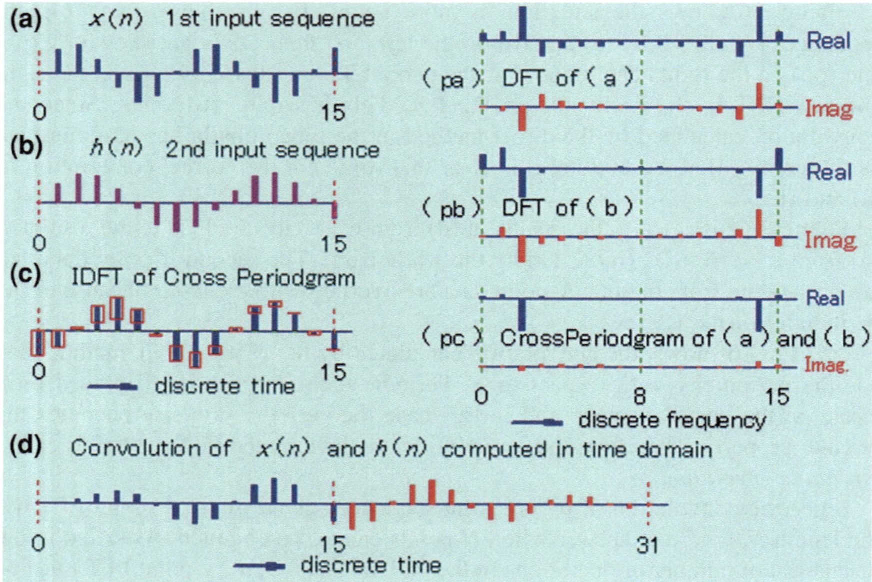

Fig. 1.6 Process of convolution calculation of $x(n)$ and $h(n)$ by 16-point DFT. **a**, **b** 16-point input sequences $x(n)$ and $h(n)$, (pa), (pb): 16-point DFTs of $x(n)$ and $h(n)$ (*top* real, *bottom* imaginary), (pc): product of (pa) and (pb) (*top* real, *bottom* imaginary), **c** IDFT of (pc), **d** convolution calculated by the direct method. Animation available in supplementary files under filename E8-06_NN_Convol.exe

1.6 Calculation of Filter Output by FFT Technique

Hereafter, in most cases, the FFT will be used instead of the DFT since it is assumed that the FFT will be used for the purpose of fast computation.

In the previous section, it was made clear that the convolution of two sequences with length N is obtained by the $2N$-point FFT/IFFT. In the digital filter output calculation, the convolution is applied to a long input sample sequence $x(n)$ and one (constant) impulse response $h(n)$. The same method shown in Fig. 1.7 can be also used for this case. The difference is that the input sequence is much longer. Of course, only one FFT is necessary for the impulse response.

The procedure (steps ①–⑧) will be explained using Fig. 1.8.

① Choose N ($=2^M$, M: integer) which is larger than the length of the impulse response $h(n)$. The input sequence $x(n)$ is divided into subsequences with length N, which are named as #0, #1, …, #p, … as shown at the top of Fig. 1.8.

② Add zeros to the end of $h(n)$ and make a $2N$-point impulse response $h_{2N}(n)$. Obtain $H_{2N}(k)$ $(0 \leq k \leq 2N - 1)$ from $h_{2N}(n)$ by the FFT.

③ Repeat the process from ④ to ⑦ starting with $p = 0$.

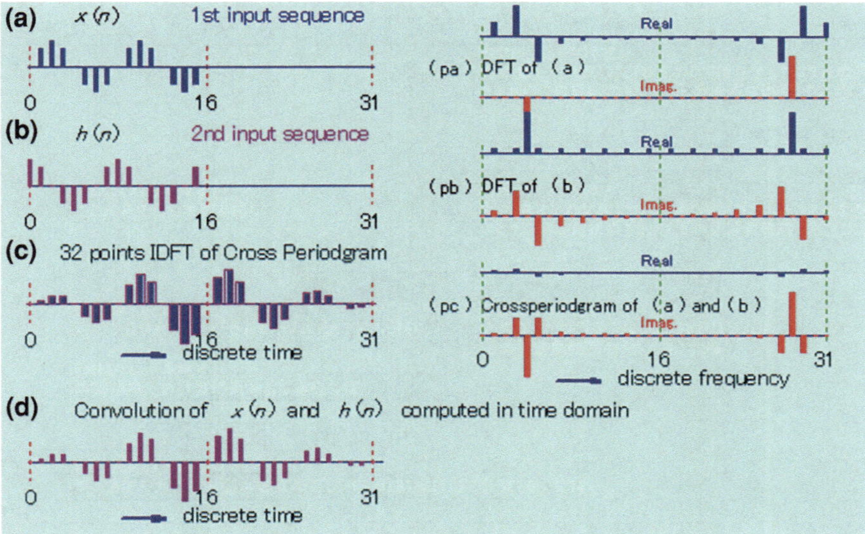

Fig. 1.7 Process of convolution calculation by 32-point DFT with the same 16-point $x(n)$ and $h(n)$ as in Fig. 1.6. **a**, **b** 32 point sequences made from 16-point input sequences $x(n)$ and $h(n)$, (pa), (pb): 32-point DFTs of (**a**) and (**b**) (*top* real, *bottom* imaginary), (pc): product of (pa) and (pb) (*top* real, *bottom* imaginary), **c** 32-point IDFT of (pc), **d** convolution calculated by the direct method. Animation available in supplementary files under filename E8-07_NFConvolution.exe

④ Select sub-sequences #p. Add zeros to the end of each and make a 2N-point input data $x_{2N}(\#p)$. Obtain $X_{2N}(k)$ $(0 \le k \le 2N - 1)$ from $x_{2N}(\#p)$ by the FFT.

⑤ Calculate $Y_{2N}(k) = X_{2N}(k) \cdot H_{2N}(k)$ (2N multiplications of complex numbers).

⑥ Inverse FFT of $Y_{2N}(k)$ and store the result ($R(\#p)$ in the figure) as $y_p(pN + n)$.

⑦ If input data is left over, add 1 to p and go back to ④. If no input data is left over, go to ⑧.

⑧ Add all $y_p(pN + n')$ $(0 \le n' \le 2N - 1)$ to obtain $y(n)$ $(n = pN + n')$ starting from $p = 0$. The addition of $y_p(pN + n)$ can be done as each of them is calculated, instead of adding them all at once at the end.

Let's compare numerical results of convolution calculations by the direct time domain method with the FFT method described here. Those are shown in Fig. 1.9. Data shown in Fig. 1.9a, b, and c are the impulse response, input sequence, and the result of the convolution calculated by the use of Equations (1.1) or (1.2) in the time domain. Data shown in Fig. 1.9d is the result of convolution obtained by steps from ④ to ⑦ for $p = 0$. Data given in Fig. 1.9e is the result obtained by step ⑧. The plot in Fig. 1.9f is the difference between (c) and (f), which is completely zero indicating that the two results agree.

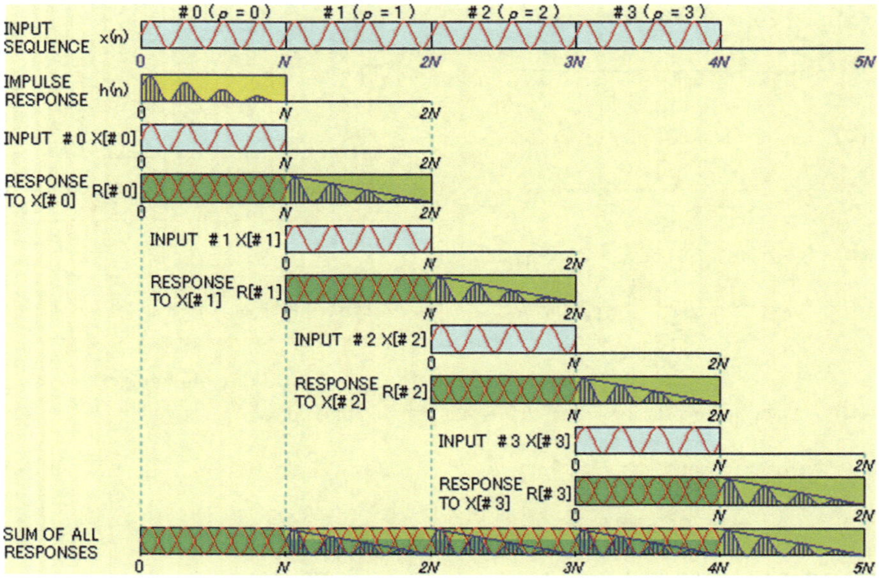

Fig. 1.8 Data handling in the convolution calculation by the FFT method

The above example is for the case with $N = 256$. Cases with different impulse responses and input sequences can be studied using the program attached to Fig. 1.9 in the CD.

1.7 Determination of Filter Coefficients from Frequency Responses

The method of convolution described in the previous chapters can be applied to any two finite sequences, which do not need to be those of a real transfer system function. Given an imaginary transfer system function with an impulse response that satisfies purpose of analysis, then the output of that system to an arbitrary input can be calculated using convolution. In this chapter, a method of determining the filter coefficients (impulse response) of a filter that satisfies any desired frequency response will be examined.

First of all, consider a *band-pass filter* (BPF) response, which is a transfer system function that passes components of the input in a limited frequency band and rejects those in all other bands.

The reader may think that the filter coefficients can be obtained from the IFFT of the transfer function (i.e., the impulse response of the system) of the filter. But it

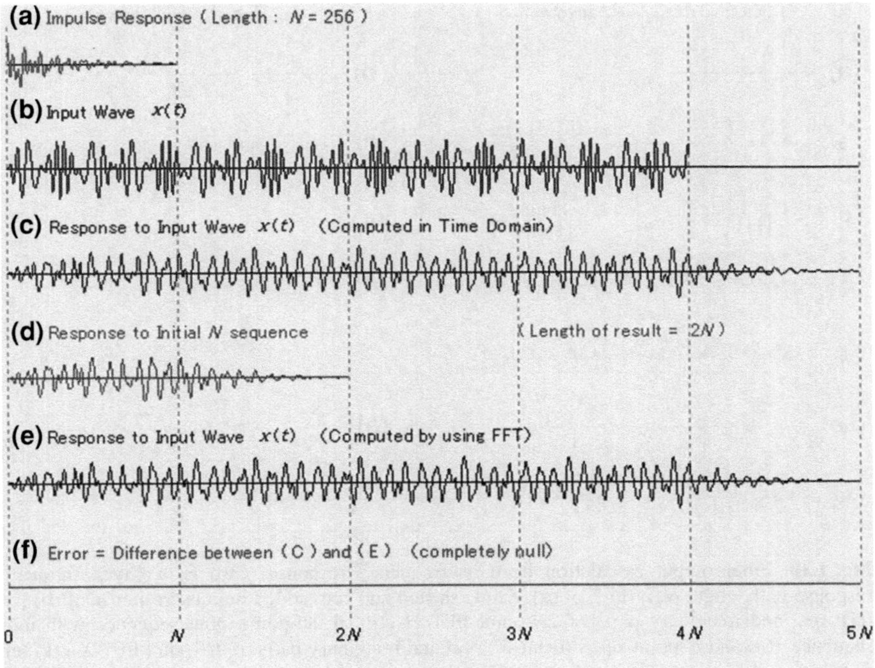

Fig. 1.9 Example of convolution calculation for a *N*-point impulse response and a long input sequence (*N*: 256). Animation available in supplementary files under filename E8-09_FFT-Convolution.exe

must be remembered that the circularity of the circular convolution must be avoided.

The amplitude characteristic (frequency response) of the band-pass filter is defined as follows: the output amplitude is equal to zero for $f < f_1$, equal to unity for $f_1 \leq f \leq f_2$, and equal to zero for $f > f_2$. The absolute value of the transfer function $H(f)$ is given by

$$H(f) = \begin{cases} 1 & f_1 \leq f \leq f_2 \\ 0 & f < f_1, \quad f > f_2 \end{cases} \tag{1.19}$$

This is the ideal band-pass frequency response, which is shown in Fig. 1.10a.

For the figures in this section, the discrete frequencies are shown in the order from 0 to N. Therefore, in Fig. 1.10a, the left and the right ends of the abscissa are for $k = 0$ and $N - 1$, and the left and right halves of the abscissa are for positive and negative frequencies, respectively. The maximum frequency f_x is at $k = N/2$, but this could be considered either as a positive or a negative frequency.

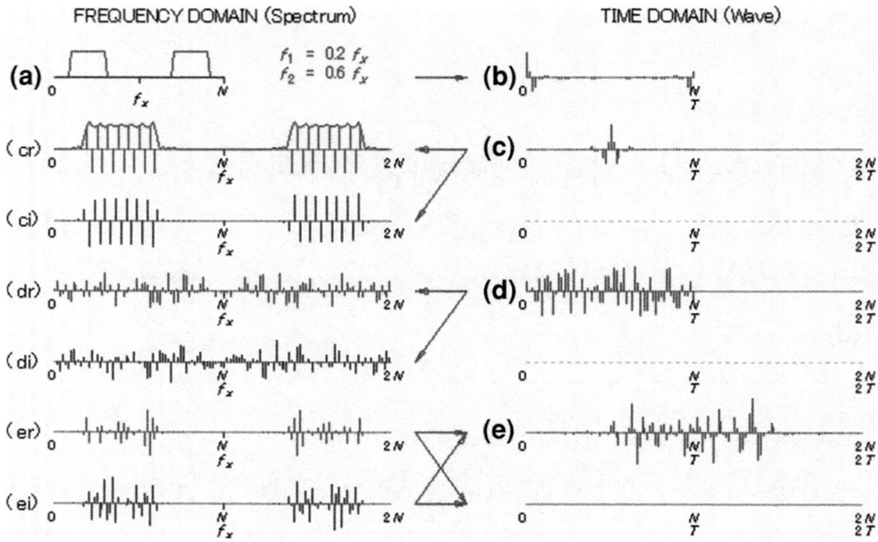

Fig. 1.10 Filter output calculation from its frequency response (case I). **a** Given frequency response with length N, **b** IFFT of (**a**), **c** time shifted and zero added impulse response of (**b**) (cr), (ci): real and imaginary parts of $2N$-point FFT $H_{2N}(k)$, **d** $2N$-point input sequence with input sequence $x(n)$ and N-point zeros (dr), (di): real and imaginary parts of $2N$-point FFT $X_{2N}(k)$ (er), (ei): real and imaginary parts of $X_{2N}(k)H_{2N}(k)$, **e** real and imaginary parts of $2N$-point IFFT of $X_{2N}(k)H_{2N}(k)$. Animation available in supplementary files under filename E8-10_BP-FILTER.exe

The low and high cut-off frequencies of the BPF are set at 0.2 f_x and 0.6 f_x, respectively. Therefore, $H(k) = 1$ for $0.1\ N \le k \le 0.3\ N$ and $0.7\ N \le k \le 0.9\ N$ and $H(k) = 0$ for $0 \le k < 0.1\ N$, $0.3\ N < k < 0.7\ N$ and $0.9\ \mathrm{N} < k < N$. The response is shown in Fig. 1.10a.

The impulse response is obtained by the N-point IFFT of $H(k)$, which is shown in Fig. 1.10b. Since the frequency response is an even function, its IFFT is purely real. The length of the response is equal to N times the sampling time, which is denoted by T at the right end of the abscissa.

Note the unexpected fact that the impulse response takes large values near the end of the response time. This is due to the circularity of the FFT. The positive and negative time regions are $1 \le n < N/2$ and $N/2 < n \le N - 1$, respectively. When considered this way, it can be seen that T is the total length of the response but not the end of the response time. The left half of the abscissa including zero $0 \le t < T/2$ is for the positive time region and $-T/2 \le t < 0$ is for the negative time region.

In Fig. 1.10c, the positive and negative regions have been swapped and $t = 0$ is set at the center of the axis. This is shown on the left half of the axis $0 < n \le N$. If it is kept this way, the impulse response seems to start from a negative time, $t = -T/2$, which is against the principle of causality. But, by shifting the time axis so that the left end corresponds to $t = 0$, this problem is removed. Now, the time needed for the response to take its maximum value is $T/2$. This is the delay of the signal incurred by going through the filter, which is called the *group delay*.

So far, the method to obtain the coefficients of the digital filter from the frequency response of the transfer function has been made clear. That is, if the left half of (c) is used as the coefficients of the FIR filter, the output of the filter is the convolution calculated in the time domain, according to Eqs. (1.1) or (1.2). However, when calculated in the frequency domain, if the impulse response length is kept equal to N, an error due to circularity is induced. To avoid this problem, N-point zeros shown by the right half of (c) must be added and the total response length must be made equal to $2N$. Fig. 1.10c can be used as the impulse response.

The $2N$-point FFT of Fig. 1.10c is the transfer function $H_{2N}(k)$ used for the calculation in the frequency domain. The real and imaginary parts of $H_{2N}(k)$ are shown by Fig. 1.10 (cr) and (ci), respectively. Their lengths are both equal to $2N$, since each is obtained by $2N$-point FFT. Since the length of (c) is $2T$, the sample period in the frequency domain is $1/2T$, which is 1/2 of (a). The abscissas of (cr) and (ci) represent the same frequency range of (a) using twice the amount of data. Then the maximum frequency f_x is at $k = N$, the center of the axis.

The reader may have noticed that (cr) and (ci) have alternating positive and negative line spectra. This is due to the phase shift induced by the time delay, $T/2$ from (b) to (c). The curve shown in (cr) is the amplitude of the transfer function ($\sqrt{cr^2 + ci^2}$), which is slightly wavy. The reason for this will be discussed shortly.

The input is a computer-generated random sequence. This must be made a $2N$-point sequence by adding N-point zeros, which is shown in (d). The real and imaginary parts of the $2N$-point FFT of (d) are shown in (dr) and (di).

The response in the frequency domain is obtained by $Y_{2N}(k) = X_{2N}(k)H_{2N}(k)$ using [(cr), (ci)] and [(dr), (di)]. The real and imaginary parts of $Y_{2N}(k)$ are shown in (er) and (ei), respectively, at the left bottom of the figure. The reason why the frequency components except for those in the pass band are missing is that (cr) and (ci) have nonzero components only in the pass band.

The $2N$-point IFFT, with (er) and (ei) as the real and imaginary parts, is shown by (e), which of course agrees with the convolution of (c) and (d) in the time domain. That the time when the large response appears is significantly delayed from $n = 0$ is due to the group delay of the filter. If the input sequence is longer than N, the procedure described using Figs. 1.8 and 1.9 can be used.

It has been made clear what happens to a signal when it goes through a transfer system with a given frequency response.

The above discussion raises an important consideration, which should be kept in mind. Suppose one wants to get a waveform with a band-limited spectrum. The correct answer will not be achieved by applying an N-point FFT to an original waveform, band-limiting the spectrum (i.e., letting some components be zero), and then applying an N-point IFFT to it. Instead, N-point zeros must first be added to the original waveform and then the procedure shown in this section must be followed. However, if one wants to obtain the energy in a frequency band (integration of the power in the time window), those extra steps are not needed. Simply

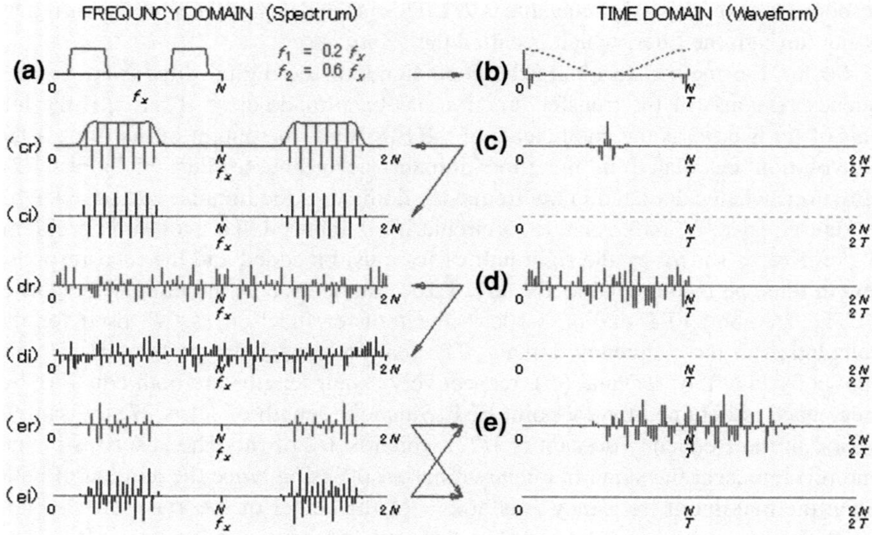

Fig. 1.11 Filter output calculation from its frequency response (case II). **a** Given frequency response with length *N*, **b** IFFT of (**a**) weighted with the function shown by the *dotted line*, **c** time shifted and zero added impulse response of (**b**) (length 2*N*) (cr), (ci): real and imaginary parts of 2*N*-point FFT $H_{2N}(k)$, **d** 2*N*-point input sequence with input sequence $x(n)$ and *N*-point zeros (dr), (di): real and imaginary parts of 2*N*-point FFT $X_{2N}(k)$ (er), (ei): real and imaginary parts of $X_{2N}(k)H_{2N}(k)$, **e** output obtained by 2*N*-point IFFT of $X_{2N}(k)H_{2N}(k)$. Animation available in supplementary files under filename E8-11_BP-FILTER.exe

apply the *N*-point FFT and add the power spectra (sum of squares of the real and imaginary parts) in the band.

Next, let's attack the remaining problem. The thin line in Fig. 1.10 (cr) shows the absolute value of the transfer function. As was mentioned before, it has small fluctuations. Our present task is to solve this problem.

The reason for the fluctuation is that the impulse response Fig. 1.10c has small but abrupt changes (from zero to a finite value and vice versa) at the beginning and the end of the response. The same problem was encountered in Chap. 6. From that experience it is known that the problem may be solved by smoothing both ends of the response, Fig. 1.10c, using the Hanning or other similar windows. Results are shown in Fig. 1.11. Figure 1.11a is the same as Fig. 1.10a. Figure 1.11b is the *N*-point FFT of (a) weighted by the Hanning window (shown by the dotted line).

$$W_{\mathrm{H}}(n) = 0.5 + 0.5\cos(2\pi n/N)$$

Figure 1.11c is obtained by the time reversal and time shift of Fig. 1.11b by the same way as Fig. 1.10c was obtained from Fig. 1.10b. The first half of the *N*-point sequence in Fig. 1.11c smoothly approaches zero at both ends.

The real and imaginary parts of the 2*N*-point FFT of Fig. 1.11c are shown by (cr) and (ci), respectively. The thin curves in Fig. 1.11 (cr) show the envelope of the transfer function. In the pass band, it is very flat indicating that the problem has been solved.

The method of determining the filter coefficients described in this chapter can be applied not only to the BPF but also to any type of frequency response.

1.8 Exercise

1. Derive Eq. (1.1) from Eq. (1.2).
2. The length of $x(n)$ is N and the length of $y(n)$ is M. What, then is the length of $x(n) * y(n)$?
3. The length of $x(n)$ is N and the length of $y(n)$ is M. If $N > M$, how many multiplication and summation operations per sample of $x(n) * y(n)$ are necessary?
4. $X(k)$ and $H(k)$ are the N-point FFTs of $x(n) = \sin(2\pi pn/N)$ and $h(n) = \sin(2\pi mn/N)$, respectively. Discuss results of $X(k)H(k)$ for the cases with $p \neq m$ and $p = m$ (p and m are both integers).
5. Sequences $x(n)$ and $h(n)$ are equal to $\sin(2\pi pn/N)$ and $\sin(2\pi mn/N)$ for $0 \leq n \leq N-1$, respectively, and both sequences have zero data for the remaining samples. Discuss the result of the convolution of $x(n) * h(n)$ for the cases with $p \neq m$ and $p = m$ (p and m are both integers).
6. Sequences $x(n)$ and $h(n)$ take nonzero values for $0 \leq n \leq N-1$ and $0 \leq n \leq M-1$, respectively, and both sequences have zero data for the remaining samples. If you want to compute $x(n) * h(n)$ using the shortest sequences, what should be the sequence length?
7. In order to compute a convolution of a longer sequence $x(n)$ and a shorter sequence $h(n)$ by the same method as shown in Fig. 1.8, a 2*N*-point FFT method is used. Discuss the procedure for the case when an 8*N*-point FFT method is more appropriate.
8. What is the meaning of "convolution in the frequency domain"?
9. Discuss the advantages and disadvantages of calculations of convolution in the time and frequency domains.

Chapter 2
Correlation

Correlation is a function in the time-delay domain that represents a possible relationship between two functions. A signal analysis technique that uses the correlation function is referred to as "correlation analysis," a technique that was commonly used for signal analysis until the 1960s. But correlation analysis has been pushed to a backseat role since the appearance of the FFT technique. This is because the cross- (and auto-) correlation and cross- (and auto-) power spectra forms a Fourier transform pair, and the correlation functions are calculated much faster from the cross- and auto-spectra by using the Fast Fourier Transform.

However, understanding of the correlation function is very important as the basis of signal analysis, even in an era when the DFT is the main tool of modern signal analysis.

2.1 Similarity of Two Sequences of Numbers

To make the meaning of the correlation clear, let us consider the similarity between two sequences of numbers, $x(n)$ and $y(n)$. This will in due course lead to the derivation of a formula for the cross-correlation function.

The sequences $x(n)$ and $y(n)$ need not be time signals, however, since the primary interest is in time signal analysis, we assume that they are samples of time signals. They can be either real or artificial. The number n represents the discrete time and therefore it is treated as the time itself.

If $x(n)$ and $y(n)$ are identical, the similarity is the greatest. In order to quantify the degree of similarity it will suffice to calculate squares of differences of the two functions at corresponding times and add them up.

Even if the two waveforms are the same but the magnitudes are different, there are differences between them. If the numerical values are large, the sum of the squares of the difference will be large and vice versa. Therefore, a normalization procedure is necessary so that the result is not affected by the magnitudes of the waveforms.

K. Kido, *Digital Fourier Analysis: Advanced Techniques*,
DOI: 10.1007/978-1-4939-1127-1_2,

For the normalization, the sequences should be divided by their respective standard deviations. The standard deviation of a sequence is actually its effective amplitude, i.e., the root-mean-square of the power of the alternating component. The division by the standard deviations removes the differences of magnitudes of the waveforms. We are more interested in the shapes of the waveforms.

If the signals contain constants (i.e., dc components), they can be removed and the similarity of the "changes of the waveforms" can be discussed.

Consider two sequences with zero means and normalized by respective standard deviations. The difference between the two sequences can be represented by the sum of the squares of the differences of values at corresponding times. If one of the sequence is shifted on the time axis, the difference (or, conversely, similarity) of the two sequences will change. This dependence of the similarity on the amount of the time shift is important information. This time shift is referred to as *(time) lag*. The similarity is therefore a function of the time lag.

In order to normalize the sequences, $x(n)$ and $y(n)$, their standard deviations, σ_x and σ_y, must be calculated. The standard deviations with sequence length L are defined by Eqs. (2.1) and (2.2).

$$\sigma_x = \sqrt{\frac{1}{L}\sum_{n=0}^{L-1}\{x(n) - \bar{x}\}^2} \tag{2.1}$$

$$\sigma_y = \sqrt{\frac{1}{L}\sum_{n=0}^{L-1}\{y(n) - \bar{y}\}^2} \tag{2.2}$$

where $\bar{x}$, $\bar{y}$ are means of $\{x(n)\}$ and $\{y(n)\}$, respectively. At present, it is assumed that the means are zero for simplicity.

The sum of squares of the differences of the sequence $x(n)$ and m-time- shifted sequence $y(m+n)$, both normalized by the individual standard deviations, is given by

$$e_m^2 = \frac{1}{L}\sum_{n=0}^{L-1}\left(\frac{x(n)}{\sigma_x} - \frac{y(m+n)}{\sigma_y}\right)^2 \tag{2.3}$$

This is modified as

$$\begin{aligned} e_m^2 &= \frac{1}{L}\sum_{n=0}^{L-1}\left(\frac{x(n)^2}{\sigma_x^2} + \frac{y(m+n)^2}{\sigma_y^2} - 2\frac{x(n)y(m+n)}{\sigma_x\sigma_y}\right) \\ &= 2\left(1 - \frac{1}{L\sigma_x\sigma_y}\sum_{n=0}^{L-1}x(n)y(m+n)\right) \end{aligned} \tag{2.4}$$

The value of this equation is equal to 0 when normalized $x(n)$ and $y(m + n)$ are identical. Therefore, it better satisfies our intention to calculate the inverse degree of "similarity" of the two sequences.

2.2 Cross-Correlation Function

Equation (2.4) represents the "degree of similarity" between the two functions. Considering only the case, $m = 0$, if the two functions are identical, it is equal to 0; if there is no relation between them, it is equal to 2; and if they have the same magnitudes and the signs are opposite, it is equal to 4. Thus, it is not necessary to use the whole of Eq. (2.4). The second term may be more suitable since it becomes 1 when the two sequences are identical, 0 when they are random, and −1 when they have the identical magnitude but opposite signs.

The second term of Eq. (2.4) can be used to represent the degree of similarity of the two functions.

$$r_{xy}(m) = \frac{1}{L\sigma_x\sigma_y}\sum_{n=0}^{L-1} x(n)y(n+m) \tag{2.5}$$

$$= \frac{\sum_{n=0}^{L-1} x(n)y(n+m)}{\sqrt{\sum_{n=0}^{N-1} x^2(n)}\sqrt{\sum_{n=0}^{N-1} y^2(n)}} \tag{2.6}$$

This is referred to as the *cross-correlation function*. As the above equations show, the correlation function is a function of lag m.

The convolution equations, Eqs. (1.1) and (1.2), and the cross-correlation equations, Eqs. (2.5) and (2.6), are almost the same. The difference is that convolution is a function of n obtained by adding terms $x(n-p)$ $h(p)$ for all p, while correlation is a function of m obtained by adding terms $x(n)$ $y(m + n)$ for all n. This is illustrated in Fig. 2.1.

First, consider the convolution function. The output of a system at a discrete time n is the addition of the products of

(1) $x(n)$, the input at time n, and the impulse response $h(0)$,
(2) $x(n-1)$, the input at one sample time before n, and the impulse response $h(1)$,
...
(3) $x(n-p)$, the input at p sample time before n, and the impulse response $h(p)$,
...

Figure 2.1a illustrates this process. The time-reversed impulse response $h(k)$ is displayed so that $k = 0$ corresponds to the discrete time n. Then, the response of

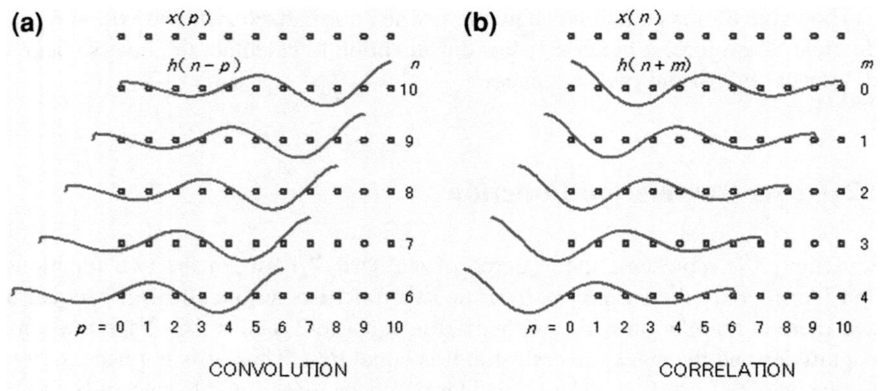

Fig. 2.1 Comparison of handling of data between the convolution and the cross-correlation calculations

the system $y(n)$ is obtained by adding the products of each of $x(p)$ (shown by the top row) and its corresponding term $h(n\text{-}p)$ (shown by the second row for $n = 10$, the third row for $n = 9, \ldots$).

In contrast, as Fig. 2.1b shows, the cross-correlation between $x(n)$ and $h(n)$ is obtained by multiplying each $x(n)$ with its corresponding $h(n)$ with the latter forward-shifted by m samples (shown by the $(m-1)$-th row in (b)) and adding all the product terms for each m.

As Fig. 2.1 shows, the convolution of $x(n)$ and $h(n)$ can be obtained as the cross-correlation of the two functions when one of the two functions is time reversed.

In Eq. (2.6), the correlation operation is defined between two sequences with the finite length L. If this number is small, correlation shows a property of limited portions of the two sequences. In the derivation of Eq. (2.5), continuous data of length L are used from $n = 0$ to $L-1$. In this case, however, the length L must be very large to represent the whole length of the sequences. Instead of using the continuous data as in Eq. (2.5), the cross-correlation can be defined in the form of an ensemble mean shown below.[1]

$$r_{xy}(m) = \frac{1}{\sigma_x \sigma_y} < x_n y_{n+m} > \tag{2.7}$$

[1] If the number of observed data is large enough to represent the property of the parent population, separations among data do not have to satisfy the sampling theorem. However, m must satisfy the criterion of the sampling theorem.

From the standpoint of signal analysis, the correlation of signals in a short period that satisfies a specific condition is sometimes more useful. It is also more convenient if the length L is finite for the practical reason of calculation. If analysis of a long time range is necessary, the equation below is useful, which makes it possible to keep the number of data equal to L while extending the effective analysis length equal to PL.

$$r_{xy}(m) = \frac{1}{L\sigma_x\sigma_y}\sum_{n=0}^{L-1} x(nP)y(nP+m) \tag{2.8}$$

n in Eq. (2.5) is multiplied by P, which means analysis range L is P-times extended, but the correlation function is correctly calculated so long as the sampling period is unchanged. There is a chance that the result is affected by the properties of contaminating noises related to a specific value of P. If this is the case, the value of P should be randomly varied.

The reader should understand from the derivation of the cross-correlation function that, if $y(n)$ is generated by giving lag m to $x(n)$, the cross-correlation between them will be 1 when the lag is equal to m. Figure 2.2 shows the case with random signal $x(n)$.

The waveforms of the top two rows in Fig. 2.2 are $x(n)$ and $y(n)$ are computer simulated. The signal $y(n)$ is made from $x(n)$ by giving a 25-sample delay. It is difficult to see that they are actually the same signal until one is informed. The cross-correlations calculated by using Eq. (2.5) are shown by the bottom three rows of the figure for $L = 10$, 100, and 1,000. The reader can barely see a peak at $m = 25$ for $L = 10$. For $L = 100$, one can almost say that there is only one peak at $m = 25$. If L is made larger up to 1,000, the noisy up-downs almost disappear. Even for an ideal random sequence simulated by a computer, a large number of averages is necessary. This suggests that averaging must be quite large when the correlation is applied to actual data.

Figure 2.2 shows the results when Eq. (2.5) is directly applied to the data. The results show that the averaging of 10 sequences is too small in order to detect the correlation. Equation (2.8) may help extend the range of averaging.

The above example is a case with no external noise. Even if a noise is added to the data, the effect of the noise contamination will be reduced by increasing the number of averaging. Figure 2.3 shows a case when a random noise with nearly the same power but with no correlation with $y(n)$ is added to $y(n)$. For $L = 10$, it became more difficult to identify the peak at $m = 25$ than the previous example. However, with increased averaging number, the position of the peak is made clear.

As shown above, if two sequences contain common signals with some delay between them, the existence of a common component and the delay time can be found by use of the cross-correlation function.

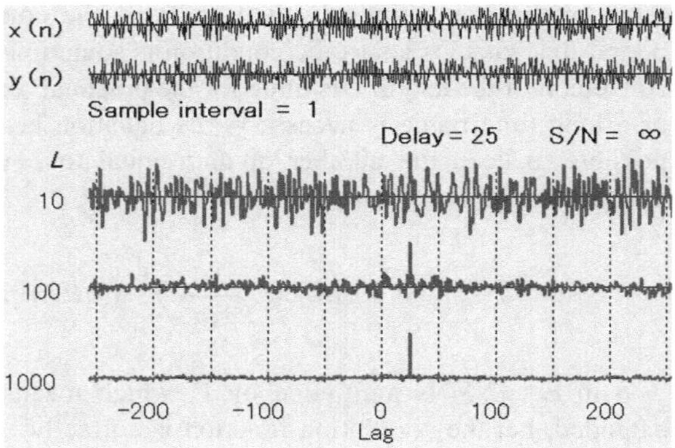

Fig. 2.2 Comparison of the cross-correlation functions of a random sequence and its 25-sample delayed sequence for averaging numbers $L = 10$, 100, and 1,000. Animation available in supplementary files under filename E9-02_Correlation.exe

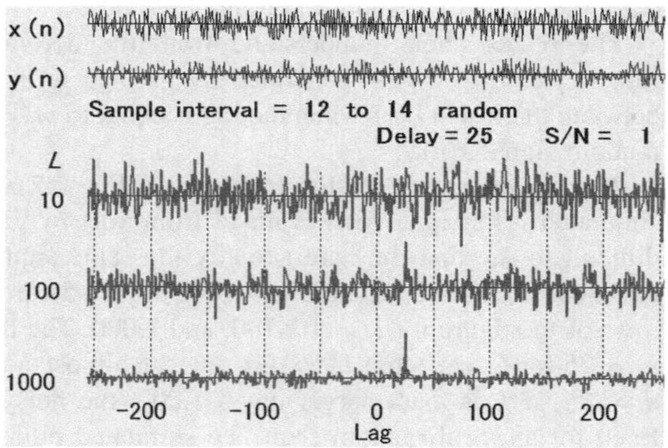

Fig. 2.3 Cross-correlation functions when a random noise with the same power but with no correlation with $x(n)$ is added to $y(n)$. The sequence $x(n)$ and $y(n)$ before the noise addition are the same as those in Fig. 2.2. Animation available in supplementary files under filename E9-03_Correlation.exe

2.3 Cross-Correlation Function Between Input and Output Signals of a Transfer System

The output $y(n)$ of a transfer system with impulse response $h(n)$ and input $x(n)$ is given by a convolution between $x(n)$ and $h(n)$. Since $y(n)$ is made from $x(n)$, there should be some kind of relationship between $x(n)$ and $y(n)$, and it should appear in the cross-correlation.

By taking the range of p from 0 to $K-1$ in Eq. (1.1), the convolution between $x(n)$ and $h(n)$ is given by

$$y(n) = \sum_{p=0}^{K-1} x(n-p)h(p) \tag{2.9}$$

Substituting this equation into Eq. (2.5), the cross correlation r_{xy} is obtained

$$r_{xy}(m) = \frac{1}{L\sigma_x\sigma_y}\sum_{n=0}^{L-1} x(n)\sum_{p=0}^{K-1} x(n+m-p)h(p)$$

Reversing the order of summation,

$$r_{xy}(m) = \frac{1}{L\sigma_x\sigma_y}\sum_{p=0}^{K-1} h(p)\sum_{n=0}^{L-1} x(n)x(n+m-p) \tag{2.10}$$

is obtained. If $x(n)$ is a random sequence, $x(n)x(n+m-p)$ takes positive and negative random values except for the case $m = p$. Therefore, if L is large enough, $x(n)x(n+m-p)$ becomes negligibly small compared to the value for $m = p$,

$$\sum_{n=0}^{L-1} x(n)x(n+m-p) = \sum_{n=0}^{L-1} x(n)x(n) = L\sigma_x^2$$

Then, if $x(n)$ is a random sequence, Eq. (2.10) is given by

$$r_{xy}(m) = \frac{L\sigma_x^2}{L\sigma_x\sigma_y}h(m) = \frac{\sigma_x}{\sigma_y}h(m) \tag{2.11}$$

Equation (2.11) shows the interesting result that the correlation is proportional to the impulse response. The effect of noise contamination is not considered in the above discussion. Even if uncorrelated noise is added to the input and/or output, the same equation Eq. (2.11) is obtained. Its derivation is given in Appendix 2A.

Let us check by numerical calculations whether the cross-correlation gives the impulse response or not.

The impulse response $h(n)$ used here is a decaying sine wave, which was also used in Chap. 1. Since $x(n)$ must be a random sequence, a random number sequence simulated by computer is used. The sequence $y(n)$ is given by the convolution of $x(n)$ and $h(n)$ and then the correlation is calculated between $x(n)$ and $y(n)$. The number of averages is examined for three cases, 10, 100, and 1,000.

The results are given in Fig. 2.4. The three charts from the top show the impulse response $h(n)$, random sequence $x(n)$, and the system output $y(n)$, respectively. The bottom three charts show the cross-correlation functions for $L = 10$, 100, and 1,000, respectively. The spacing of data (P in Eq. (2.8)) is randomly varied within 12 and 14. The results show that the averaging of 100 is not enough, but if it is increased to 1,000, a waveform similar to $h(n)$ appears in y(n) from around lag $m = 50$, which was given to $y(n)$. As the number of averages is increased, the cross-correlation function gets closer to the true impulse response, indicating that Eq. (2.11) is valid.

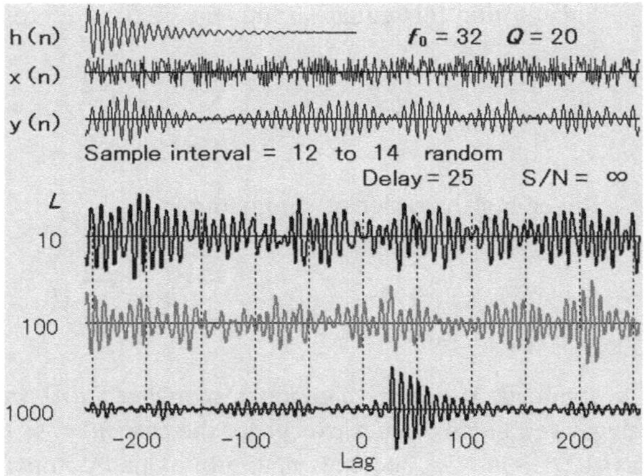

Fig. 2.4 Cross-correlations between a random input $x(n)$ and an output $y(n)$ of a transfer system with an impulse response $h(n)$ for different averaging numbers, 10, 100, and 1,000. Animation available in supplementary files under filename E9-04_Correlation.exe

More insight into cross correlation analysis can be obtained by examining the program for the data given in Fig. 2.4. It will be seen for example that the addition of noise does not deteriorate the results too much, and that a choice of data spacing P with respect to the period of $x(n)$ sometimes gives serious problems, that are not described in the text.

2.4 Auto-Correlation Function

Up to the present, relationship between two different sequences $x(n)$ and $y(n)$ has been considered. During the discussion, the special case $x(n) = y(n)$ was not excluded. In fact, if this is the case, the correlation may include some information about the sequence itself. The *auto-correlation function* is defined by replacing $y(n)$ by $x(n)$ in Eqs. (2.5) and (2.7).

$$r_{xx}(m) = \frac{1}{L\sigma_x^2}\sum_{n=0}^{L-1} x(n)x(m+n) \tag{2.12}$$

$$r_{xx}(m) = \frac{1}{\sigma_x^2} < x(n)x(m+n) > \tag{2.13}$$

It is clear that the auto-correlation function is always equal to 1 for $m = 0$. If the range of n is infinite, the two summations $\sum_n x(n)x(m+n)$ and $\sum_n x(n-m)x(n)$, i.e., $\sum_n x(n)x(-m+n)$ become equal, indicating that auto-correlation is an even function of the lag m.

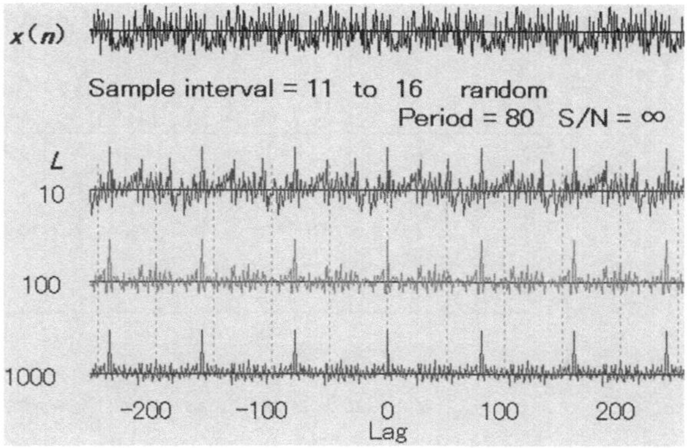

Fig. 2.5 Random sequence with period 80, $x(n)$, and its auto-correlations for averaging numbers $L = 10$, 100, 1,000. Animation available in supplementary files under filename E9-05_AutoCorrelation.exe

If the sequence is random, the auto-correlation must be zero except for $m = 0$. Random number generation functions installed in computers are almost perfect and the cross-correlation function of a random sequence generated by a computer almost always satisfies the condition that it is equal to 1 for $m = 0$ and 0 for $m \neq 0$.

Let us calculate a cross-correlation function of a periodically repeated random sequence generated by a computer. Results of the calculation are shown in Fig. 2.5 for averaging numbers $L = 10$, 100, and 1,000. Since the sequence is periodic, the auto-correlation function becomes 1 at every one period (80 points in the present case). The spacing of data (P in Eq. (2.8)) is randomly varied between 11 and 16. Figure 2.5 shows that averaging 10 times is not sufficient. It seems necessary that at least 100 averages is necessary when the sequence is random.

The above examples are auto-correlations of artificially generated, i.e., rather ideal, signals. The reader might have the impression that the results are being intentionally idealized by the author. As an example of a real signal, the auto-correlation of the number of sunspots as a function of year will be calculated. The yearly change of the number of the sunspots has been observed for approximately 300 years as shown in Fig. 2.6a. Since the number of sunspots is always positive, the mean of the auto-correlation function takes a non zero value. As a trial, let us directly apply Eq. (2.9) without making the mean equal to zero. The result is given by Fig. 2.6b. Since the auto-correlation is an even function of the lag, only the results for $m \geq 0$ are shown. Since the mean is not zero and other data except for the actually observed 294 data are made equal to zero, the auto-correlation takes larger values as the lag approaches 0 as a general trend. Still, one can read that the period of the sunspot number variation is roughly 11 years.

Figure 2.6c shows a sunspot data, which is modified so that the mean becomes equal to zero and its auto-correlation is shown by Fig. 2.6d. This seems to indicate that the sunspot variation has also a longer period, approximately 10 times that of 11 years. However, the data is too short to purport this theory.

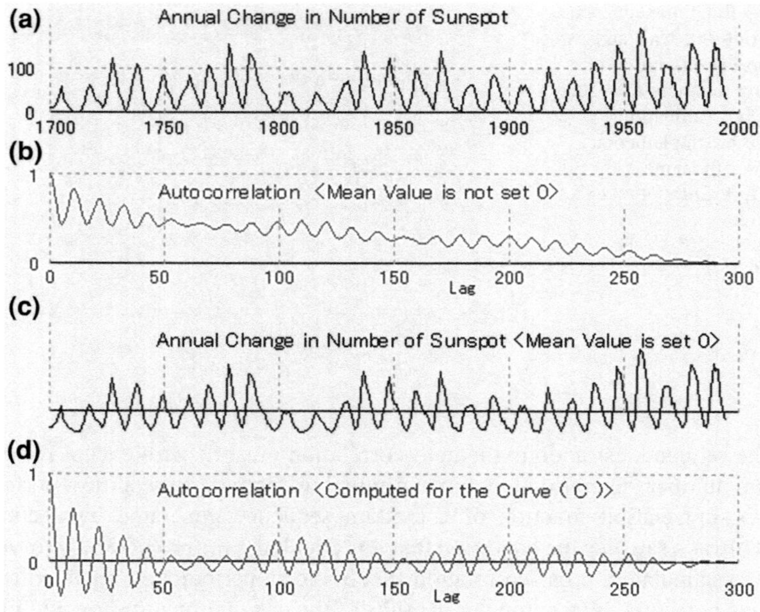

Fig. 2.6 Auto-correlation function of the yearly change of sunspots

2.5 Analysis of Sequences by Auto-Correlation Functions

In Sect. 2.4, applications of extracting periods of sequences were introduced. However, the cross- and auto-correlation functions have wider applications than this.

Let us consider an example shown in Fig. 2.7, which depicts a sound source and a wall. The sound source radiates a random noise. The sound received by the observer is a combination of the direct noise and the noise reflected from the wall. The reflected noise has some time delay compared to the direct sound. Can this time delay be determined by correlation analysis?

The waveforms are represented by sample values. If the direct sound is represented by $x(n)$, the reflected sound can be represented by $rx(n - d)$, where r is the amplitude ratio of the reflected waveform to the direct waveform and d is the time delay between them. Then, the received sound at the observation point $y(n)$ is given by

$$y(n) = x(n) + rx(n - d) \tag{2.14}$$

The auto-correlation of $y(n)$ is given by (see Appendix 2C)

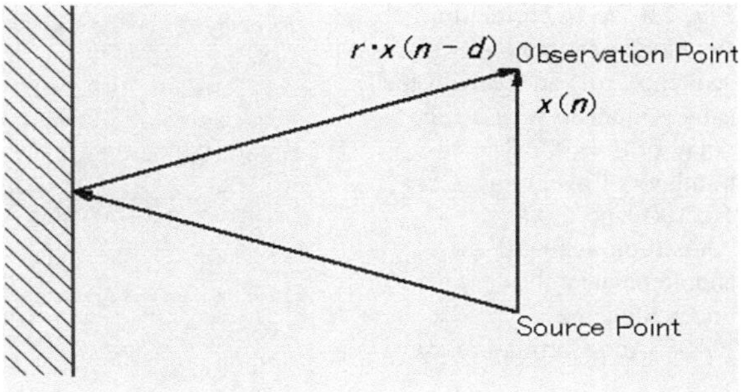

Fig. 2.7 Superposition of the direct sound $x(n)$ and the reflected sound $rx(n-d)$, with r times the amplitude and delay time d

$$\begin{aligned} R_{yy}(m) &= \frac{1}{L\sigma_y^2}\sum_{n=0}^{L-1} y(n)y(n+m) \\ &= (1+r^2)R_{xx}(m) + rR_{xx}(m+d) + rR_{xx}(m-d) \end{aligned} \tag{2.15}$$

Equation (2.15) shows that the correlation function takes the value $(1 + r^2)$ at $m = 0$, and r at $m = \pm d$. Remember that $R_{yy}(0) = 1$ and $R_{yy}(m \neq 0) = 0$ if $x(n)$ is a random sequence.

In order to confirm the above discussion, auto-correlation functions of a random sequence added to the same sequence with a time delay are shown in Fig. 2.8. The time delay corresponds to a time length of 90 samples. There are two peaks at $m = \pm 90$ with the amplitude of 1/2 of the peak at $m = 0$. This is because $r = 1$ in this case and external noise is not included.

Next, consider the auto-correlation of the observed output $y(n)$ of a system with an impulse response $h(n)$ and a random input sequence $x(n)$. This corresponds to the case of cross-correlation function of Fig. 2.4. The results are shown in Fig. 2.9 without a rather lengthy theory.

The first and second charts in Fig. 2.9 shows the output $y(n)$, the convolution of $x(n)$ and $h(n)$, and the impulse response, $h(n)$, respectively. The bottom three charts are auto-correlations for three numbers of averaging, $L = 10$, 100, and 1,000. The data spacing (P in Eq. (2.8)) is randomly chosen from 12 to 20. The impulse response given by the auto-correlation function extends to both positive and negative time directions contrary to the case of the cross-correlation function. This is because the auto correlation is an even function.

Some examples of auto-correlations of waveforms are shown in Fig. 2.10. Some of these will be referred to in later sections.

Figure 2.10a is an auto-correlation of a random sequence. As stated before, it should become 1 at 0 lag and 0 at other lags. However, even with 1,000 averages, it is not perfectly zero for non zero lags.

Figure 2.10b is an auto-correlation of a sequence obtained from a moving-average of the sequence (a) over 64 points ($x'(n) = \frac{1}{64}\sum_{m=-32}^{31} x(n+m)$). The

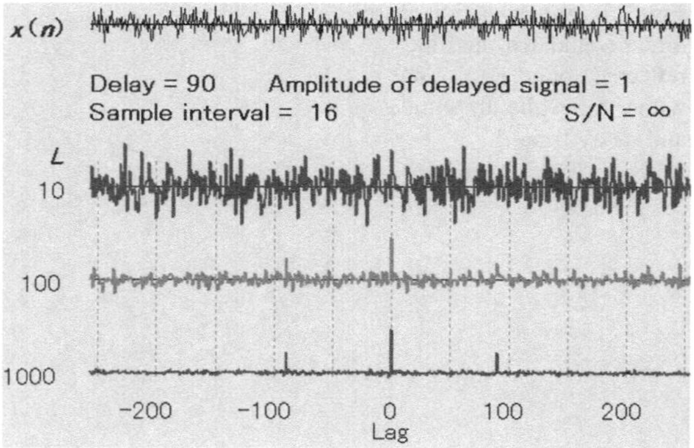

Fig. 2.8 Auto-correlation functions of a random sequence $x(n)$ added with the same sequence with a time delay of $d = 90$. The numbers of averaging L are 10, 100, and 1,000. Animation available in supplementary files under filename E9-08_AutoCorrelation.exe

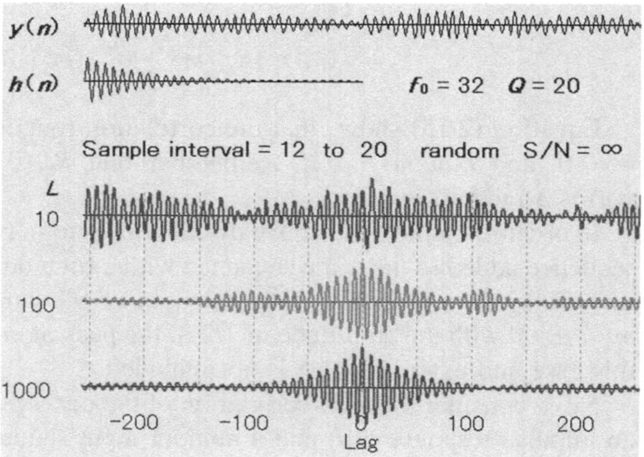

Fig. 2.9 Auto-correlation functions of $y(n)$ as an output of a system with an impulse response $h(n)$ and an input $x(n)$ (not shown). The numbers of averages L are 10, 100, and 1,000. Animation available in supplementary files under filename E9-09_AutoCorrelation.exe

auto-correlation function takes the form of an isosceles triangle with its base equal to twice the time length of the moving average.

Figure 2.10c is an auto-correlation of a sequence obtained similarly as (b) but with Hanning window weighting (see Chap. 7). The auto-correlation function is similar to the shape of the Hanning window.

Figure 2.10d is an auto-correlation of a sequence of positive and negative pulses with a random spacing from 25 to 45. The average of the sequence is equal to zero. The auto-correlation function is zero within $m = \pm 25$ (except for $m = 0$) and, outside of this range, it approaches 0 as the number of averaging increases.

Figure 2.10e is an auto-correlation of a sequence similar to the one in Fig. 2.8, but the original random sequence is a moving-average over 16 points. The auto-correlation function has a spreading similar to Fig. 2.10b.

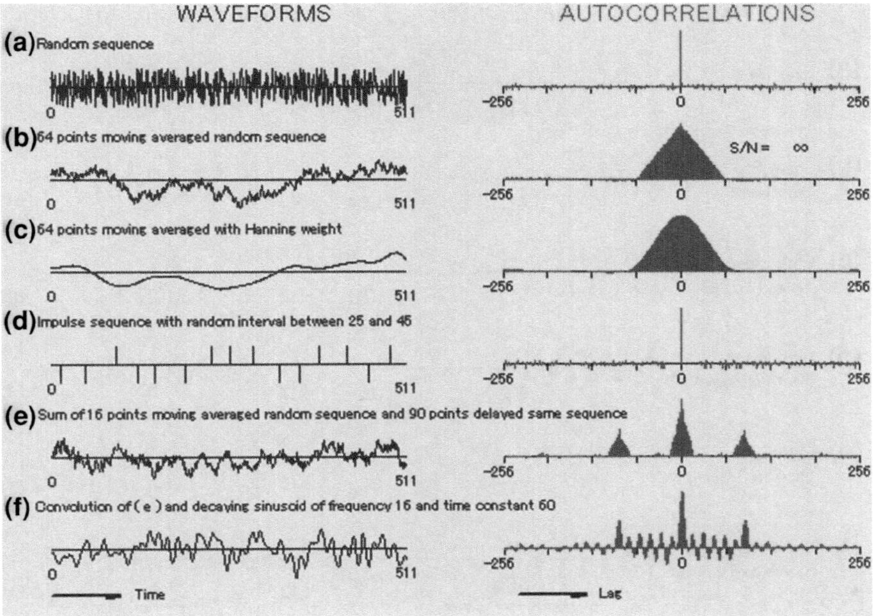

Fig. 2.10 Auto-correlation functions of several waveforms. Animation available in supplementary files under filename E9-10_AutoCorrelation.exe

Figure 2.10f is an auto-correlation of a sequence similar to the one in Fig. 2.9. Since the input sequence is Fig. 2.10e, there are many small triangular peaks around the main three peaks of Fig. 2.10e.

In the program of Fig. 2.10, the reader can check auto-correlations under many other conditions.

In the examples of Figs. 2.8, 2.9 and 2.10, noise is not considered. If an uncorrelated noise is added to the sequences, auto-correlations are decreased except for the lag at $m = 0$. One can also check these aspects in the programs of Figs. 2.8, 2.9 and 2.10.

2.6 Short-Term Auto-Correlation Function

In the previous discussions auto-correlations that represent properties over whole ranges of rather long sequences have been considered. The auto-correlation can be applied to signals that change their waveforms continuously with time, such as speech signals. In those signals, auto-correlations over long ranges of lags are not appropriate. In speech signals, a pitch can last several milliseconds and it can change in several tens to hundreds of milliseconds. The auto-correlation includes

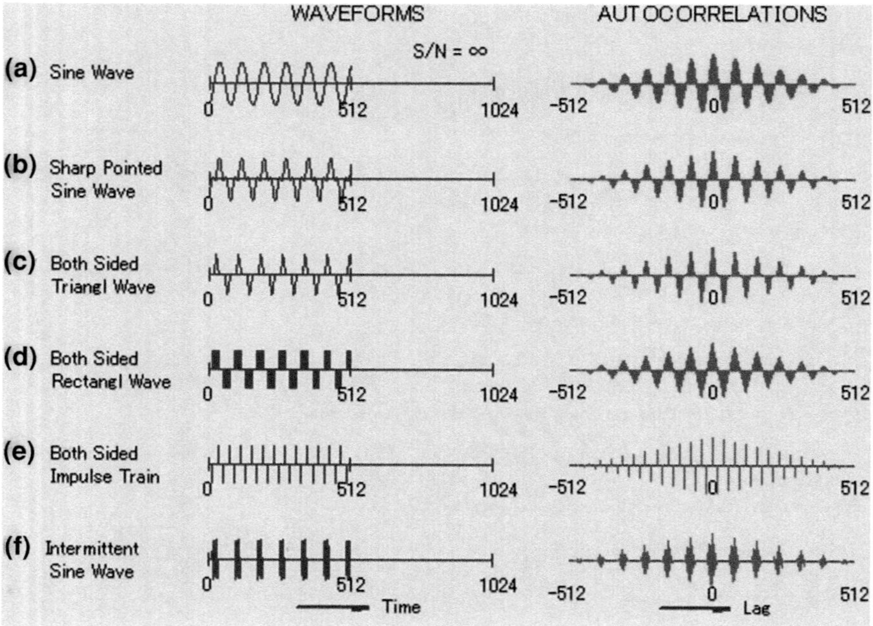

Fig. 2.11 Six periodic waveforms with 512 sequence lengths and their short-term auto-correlation functions with $L = 1024$. Animation available in supplementary files under filename E9-11_Autocorrelation.exe

not only pitch information, but also spectrum information and the latter also changes within a short duration. If the auto-correlation is used for speech analysis, it must be applied to signals with short duration, or to short portions of long signals.

Equation (2.12) can still be applied in those cases where the data length is short and the number of averages (L) cannot be made larger than twice the data length. Let's synthesize several waveforms with short periods and calculate their *short-term auto-correlations.*

Six periodic waveforms and their auto-correlation functions are shown in Fig. 2.11. The sequence length of the signal shown in the figure is 512, and outside of the sequence the signal is considered to be zero. The number of averages L, is 1,024, and therefore, the auto-correlation function reduces linearly from $m = 0$ to $m = \pm 512$. The auto correlation is zero for $|m| > 512$.

Figure 2.11 shows some interesting relations between the waveforms and their auto-correlations functions: (1) the auto-correlations of periodic waveforms also becomes periodic, (2) the auto-correlations of sinusoidal waveforms also become sinusoidal, (3) the auto-correlations of impulse sequences also become impulsive and so on. The periodic intermittent sine wave of Fig. 2.11f is a succession of the

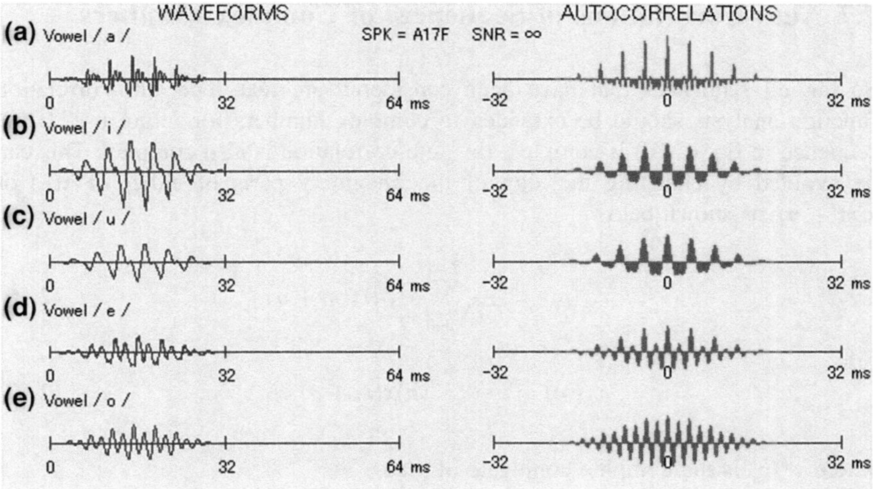

Fig. 2.12 Auto-correlation functions of five vowels with Hanning windows. Animation available in supplementary files under filename E9-12_VowelCorrelation.exe

same short sine waves. Even if each short sine wave is replaced with other types of waves, similar results will be obtained. One can obtain other findings by running the program for Fig. 2.11.

The auto-correlation function is often used to detect pitch frequencies of speech. The left-hand side of Fig. 2.12 shows waveforms of /a/, /i/, /u/, /e/, and /o/ sampled with a 16 kHz sampling frequency through the Hanning window. The first 512 points (32 ms) contain the speech waveforms and 512-point zeros are added to them. The right-hand side shows their auto-correlation functions. Since the waveforms approach zero near the ends due to the Hanning window, the auto-correlation functions quickly reduce as the lag increases in the positive and negative directions.

Since the vowels have periodic waveforms, their auto-correlation functions are also periodic. The largest peak with the minimum lag (except for the zero lag) gives the period of the pitch. They can be easily read from Fig. 2.12a–d, but that of Fig. 2.12e is not. Nevertheless, the largest peak (except that of the origin) gives the period of the pitch.

In the program associated with the analysis of Fig. 2.12, one can see the dependence of the auto-correlations functions of voices on their speakers as well as on the gender of their speakers.

2.7 Auto-Correlations of Sequences of Complex Numbers

So far, all sequences that have been considered are real. The auto-correlation function analysis should be extended to complex numbers for future use. If the sequence in Eq. (2.10) is complex, the auto-correlation is also complex. This can be avoided by changing the sign of the imaginary parts of either of $x(n)$ or $x(m + n)$, as shown below.

$$r_{xx}(m) = \frac{1}{L\sigma_x^2}\sum_{n=0}^{L-1} x^*(n)x(m+n) \tag{2.16}$$

$$r_{xx}(m) = \frac{1}{\sigma_x^2} < x^*(n)x(m+n) > \tag{2.17}$$

where $x^*(n)$ is the complex conjugate of $x(n)$.

The auto-correlation function of a continuous function $x(t)$ is given by Eq. (2.18)

$$r_{xx}(\tau) = \frac{1}{T\sigma_x^2}\int_0^T x^*(t)x(t+\tau)dt \tag{2.18}$$

where T is the range over which the waveform exists. If the signal is infinitely long, the range of integration should be from $-\infty$ to $+\infty$. The standard deviation for the continuous system, σ_x, is given by

$$\sigma_x = \sqrt{\frac{1}{T}\int_0^T x^2(t)dt} \tag{2.19}$$

The cross-correlation is also defined here as

$$r_{xy}(m) = \frac{1}{L\sigma_x\sigma_y}\sum_{n=0}^{L-1} x^*(n)y(n+m) \tag{2.20}$$

$$r_{xy}(m) = \frac{1}{\sigma_x\sigma_y} < x^*(n)y(n+m) > \tag{2.21}$$

The cross-correlation function of continuous functions $x(t)$ and $y(t)$, $r_{xy}(\tau)$, is given by

$$r_{xy}(\tau) = \frac{1}{T\sigma_x\sigma_y}\int_0^T x^*(t)y(t+\tau)dt \tag{2.22}$$

The definitions given above does not cause any change as long as sequences or functions are real. If $x(n)$ and $y(n)$ are complex, the cross-correlation defined by Eqs. (2.20) or (2.21) or (2.22) does not always become real. However, the reader will understand the advantages of the definitions given above when we address the Fourier transforms of auto- and cross-correlation functions.

2.8 Fourier Transforms of Auto-Correlation Functions

Let us calculate the Fourier transform of Eq. (2.18) with the range of integration from $-\infty$ to $+\infty$. It is assumed that $x(t) = 0$ for $|t| > Tx$, where Tx is a sufficiently large positive constant and $x^2(t)$ is integrable. We also neglect the division by T in Eq. (2.18).

$$\mathrm{FT}[r_{xx}(\tau)] = \frac{1}{\sigma_x^2}\int_{-\infty}^{+\infty}\int_{-\infty}^{+\infty} x^*(t)x(t+\tau)dt\,\exp(-j2\pi f\tau)d\tau$$

By replacing $(t+\tau)$ by u ($\tau = u - t,\quad d\tau = du$), the above equation is rewritten as

$$\begin{aligned}\mathrm{FT}[r_{xx}(\tau)] &= \frac{1}{\sigma_x^2}\int_{-\infty}^{+\infty}\int_{-\infty}^{+\infty} x^*(t)x(u)dt\,\exp\{-j2\pi f(u-t)\}du\\ &= \frac{1}{\sigma_x^2}\int_{-\infty}^{+\infty} x^*(t)\exp(j2\pi ft)dt\int_{-\infty}^{+\infty} x(u)\exp(-j2\pi fu)\}du\end{aligned}$$

Therefore,

$$\mathrm{FT}[r_{xx}(\tau)] = R_{xx}(f) = \frac{1}{\sigma_x^2}X^*(f)X(f) \tag{2.23}$$

This equation states that the auto-correlation function and the (normalized) power spectrum are a Fourier transform pair, also known as the *Wiener–Khinchine Theorem.*

If this relation exists in the continuous system, there should be a similar relationship in the discrete system. Let us calculate the DFT of Eq. (2.16) to see whether the DFT of the auto-correlation function gives the periodogrum.

Thus far, the symbol L was used for the number of averages to calculate the auto-correlation function. In the following discussions, L is changed to N, meaning also the data number of the DFT. In the calculation of the DFT, the parameter n of $x(n)$ must take values 0, 1, 2, ..., $(N - 1)$ without skipping a value.

Then, the DFT of the auto-correlation is given by

$$R_{xx}(k) = \mathrm{DFT}[r_{xx}(m)] = \frac{1}{N\sigma_x^2}\sum_{m=0}^{N-1}\sum_{n=0}^{N-1} x^*(n)x(n+m)\exp(-j2\pi\frac{mk}{N})$$

By letting $p = n + m \quad (m = p - n)$ and replacing $x(p)$ by $x(p - N)$ (periodicity of $x(p)$), the range of p from $p = n$ to $p = (n + N + 1)$ can be changed to $p = 0$ to $p = N - 1$. Then, the above equation becomes:

$$\begin{aligned} R_{xx}(k) &= \frac{1}{N\sigma_x^2}\sum_{p=0}^{N-1}\sum_{n=0}^{N-1} x^*(n)x(p)\exp\left(-j2\pi\frac{pk}{N}\right)\exp\left(j2\pi\frac{nk}{N}\right) \\ &= \frac{1}{N\sigma_x^2}\sum_{n=0}^{N-1} x^*(n)\exp\left(j2\pi\frac{nk}{N}\right)\sum_{p=0}^{N-1} x(p)\exp(-j2\pi\frac{pk}{N}) \\ &= \frac{1}{N\sigma_x^2}X^*(k)X(k) = \frac{1}{N\sigma_x^2}|X(k)|^2 \end{aligned} \tag{2.24}$$

This is the discrete version of the Wiener–Khinchine Theorem. One reason why the coefficient $1/(N\sigma_x^2)$ is added to the right-hand side is that the auto-correlation function is normalized.

The periodogram of the auto-correlation is calculated using Eq. (2.24)

$$|X(k)|^2 = N\sigma_x^2\mathrm{DFT}[r_{xx}(m)] \tag{2.25}$$

If the auto-correlation function $r_{xx}(m)$ is given for all values of lag m from $-N/2$ to $N/2$, the power spectrum is given by the N-point DFT. Since the auto-correlation is a function of m only, and the power spectrum is independent of the data spacing (P in Eq. (2.8)) used when the auto-correlation function is obtained, Eq. (2.25) is valid independent of the method by which the auto-correlation function is obtained. If the auto-correlation function represents a property of a whole sequence, the power spectrum is also of the same whole sequence.

The IDFT of the (auto) power spectrum must be the auto-correlation function, since the latter is the DFT of the former. Therefore, the auto-correlation can be calculated with less computation steps if the power spectrum is obtained first and then it is inverse-Fourier-transformed by the IFFT. It can be expressed as

$$r_{xx}(m) = \frac{1}{N\sigma_x^2}\mathrm{IFFT}[X^*(k)X(k)] \tag{2.26}$$

However, if $X^*(k)X(k)$ is obtained from one N-point sequence, the auto-correlation function is also given as that of an N-point sequence. In order to obtain $X^*(k)X(k)$ of a whole sequence, N-point power spectra can be calculated by averaging many FFT results with different starting points. But it has been explained in Chap. 6 in Volume 1 that this does not produce a correct power spectrum.

If the auto-correlation $r_{xx}(m)$ is known for lag $|m| \leq M$ and 0 for $|m| > M$, and if $2M < N$, then the N-point DFT of $r_{xx}(m)$ is exactly equal to the power spectrum of the sequence. In this case, there is no error even if it is assumed that the auto-correlation function is a periodic function with period N.

In the derivation of Eq. (2.24), the relationship that $x(p)$ for $p > N$ is equal to $x(p\text{-}N)$ was used. This assumes that $x(n)$ is periodic with period N. However, this is not correct if an N-point data from a longer sequence $x(n)$ is used. Equation (2.25), which corresponds to the Wiener–Khinchine Theorem (Eq. 2.23) of a continuous system, is valid with the condition that $x(n)$ is a periodic function. In order to calculate the auto-correlation of a long sequence by the DFT, a similar method used for the calculation of convolution must be used to avoid obtaining a circular correlation function. If $x(n)$ is an infinitely long random sequence, since the average of the products of sample values of different data numbers is zero, a circular correlation function will be not obtained. But this is the case for a known spectrum and auto-correlation, and in such a case, there is no need to calculate them.

2.9 Calculation of Auto-Correlation Functions by the FFT

In the previous section, a clue for the calculation of the auto-correlation function by the FFT method was indicated. It is clear that use of the FFT reduces the calculation load, but this process generates a circular DFT. One way to avoid this problem is to append zeros to the data. Another may be to apply appropriate windows. In any case, some degree of error may be unavoidable.

In Fig. 2.13, the handling of data in the auto-correlation calculation by the FFT method is shown. Two sequences $x(n)$ and $y(n)$ shown in the figure are identical. By doing this, it is clear which is the delayed sequence. The sizes of circles are proportional to the data numbers from 1 to 25. The two data sequences with the same sizes of circles are identical data. Figure 2.13 shows the case with a lag of four, and the diagonal lines connecting $x(n)$ and $y(n)$ become vertical if the lag is equal to zero.

In order to calculate the correlation function employing the definition, two sets of data and one with a lag 4 must be selected from a long sequence as shown in Fig. 2.13a. Then they are multiplied and added. Pairs of data used for the multiplication are $x(n)$ and $y(n + 4)$, that are connected by thin lines in the figure.

When the FFT method is used, N-point data within the two vertical dotted lines shown in Fig. 2.13b are used to calculate a periodogram. As explained before, the N-point FFT gives the Fourier coefficients of a sequence with period N. Therefore, the sequences used here are periodic as shown in (b) (in this example, data from 9 to 16 periodically appear). Then, data used by the N-point FFT method are

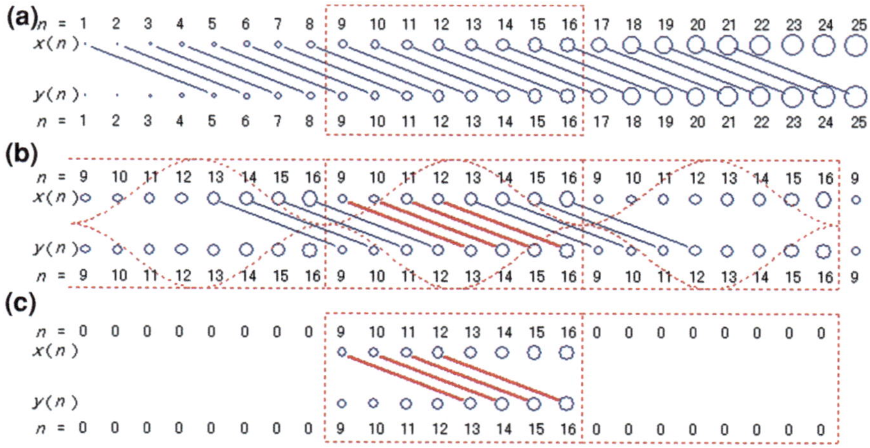

Fig. 2.13 Handling of data for the calculation of auto-correlation functions. However, the method shown by (**c**) still does not give the same result obtained by (**a**). This is because only four correct pairs indicated by *thick lines* in (**c**) are used in the calculation, compared to eight pairs in (**a**). The ratio of these correct pairs to the original pairs decreases linearly as lag *m* increases. This is the origin of the Bartlet window introduced in Chap. 7 in Volume 1

different from those in (a) except for the pairs connected by the thick diagonal lines. The pairs connected with thin diagonal lines do not exist in the original pairs shown in (a). This causes the difference between auto-correlation functions obtained by the two methods.

In the example with $m = 4$ and $N = 8$ as shown in (b), only four are original (correct) pairs and other four are wrong pairs, indicating that non-negligible errors will be induced. This type of error cannot be ignored except for cases with $m \ll N$.

It is possible to avoid the wrong pairs by adding another 8-point zero data to the data from 9 to 16 and using the 2 N-point FFT as shown in (c). Because of the circular correlation property, the data from 1 to 8 are also replaced by zeros. The "wrong" four pairs become 0 and they are excluded from the calculation by the FFT method.

So far, the rectangular window (without weighting) has been used. It was found in Chaps. 6 and 7 that the use of windows such as the Hanning window helps improve the accuracy of spectrum estimation. Our interest is in whether the Hanning window shown in Fig. 2.13b also works to improve the estimation accuracy of the correlation function. As Fig. 2.13b shows, the Hanning window is effective in reducing the values of undesirable pairs shown by thin oblique lines.

Now, four possible methods of calculating the power spectrum are available for the *N*-point sequence within the vertical dotted lines in Fig. 2.13b. The following four methods are compared.

(0) The power spectrum is obtained by the *N*-point FFT of the directly calculated correlation function.
(1) The power spectrum is obtained by the *N*-point FFT of the *N*-point sequence $x(n)$ using the rectangular window and then the auto-correlation function is calculated by the IFFT.
(2) The power spectrum is obtained by the *N*-point FFT of the *N*-point sequence $x(n)$ using the Hanning window and then the auto-correlation function is calculated by the IFFT.
(3) The power spectrum is obtained by the 2 *N*-point FFT of the *N*-point sequence $x(n)$ plus *N*-point zero data and then the auto-correlation function is calculated by the IFFT.

The source signal used here is a convolution of a random sequence and an exponentially decaying impulse response. The reason why the exponentially decaying response is used is so that we can control the lag-dependent decay rate of the auto-correlation function of the convoluted signal. If the impulse response decays quickly, the lag-dependent decay rate of the auto-correlation also decays quickly.

In Fig. 2.14, the top three charts show the random input $r(n)$, impulse response $h(n)$ and the output signal $x(n)$ (convolution of $r(n)$ and $h(n)$), respectively. The impulse response used here is synthesized by the following three steps, (1) add a 2,048 long random sequence and several sine waves, (2) reduce the amplitudes of higher frequency components of the added sequence, and (3) multiply the modified sequence by an exponentially decaying function $e^{-n/50}$ with a time constant of 50.[2]

Auto-correlation functions and power spectra calculated by the four methods are shown in the bottom four rows in Fig. 2.14. In the method (0), the auto-correlation is directly calculated following the definition using all data of $x(n)$. Since the impulse response decays with the time constant of 50, the auto-correlation decays quickly as the lag gets away from 0. The power spectrum is calculated by the FFT using 512 samples of the auto-correlation centered at m (lag) = 0.

In each of the methods from (1) to (3), the power spectrum (periodogram) is obtained as an average of 50 FFTs using data from 50 different parts of $x(n)$, and then, the auto-correlation is calculated by IFFT. Sums of squares of the differences between the auto-correlation functions obtained by the method (1)–(3) and (0) over 512 points centered at $m = 0$ are calculated and their ratio to the energy (sum of squares) of the auto-correlation function of method (0) are also shown in the figure. The method (3) using 2 *N*-point FFT has the smallest residual power

---

[2] The time constant is T of a decaying function $e^{-n/T}$, which is the time necessary for the amplitude to decay from 1 to $1/e (e : 2.7828)$.

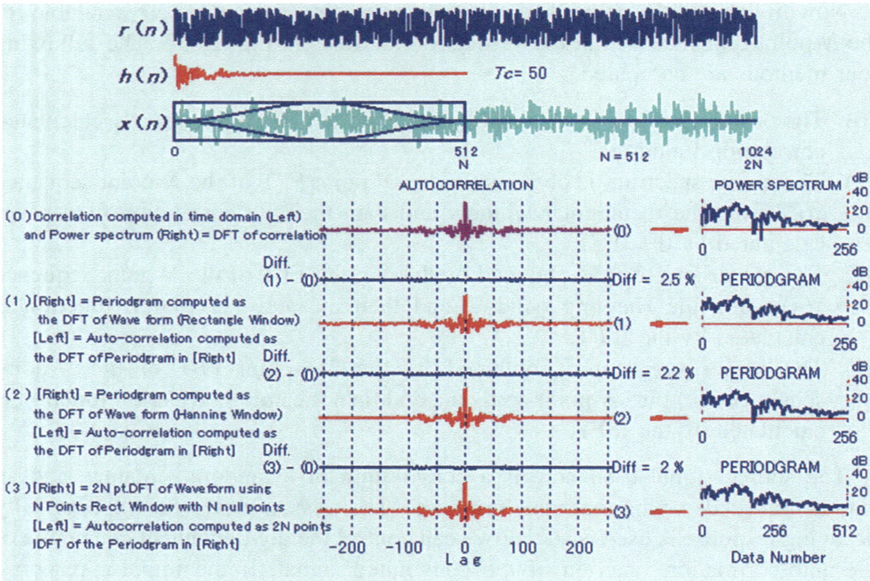

Fig. 2.14 Random input sequence $r(n)$ to a system with an impulse response $h(n)$ and the system output $x(n)$ (*top three charts*) and auto-correlations (*left*) and power spectra (*right*) calculated by the four methods from (0) to (3). Animation available in supplementary files under filename E9-14_AUTOCOR.exe

(energy) ratio, and method (1) has the largest. All three methods have the order of 2 % of the residual power ratio.

The averages of the periodic power spectra obtained by methods (1)–(3) seem to represent the power spectrum obtained by method (0) very well. This is because the spectrum of input sequence $x(n)$ is smooth and the auto-correlation decays quickly as the lag gets away from 0. In method (1), which uses the rectangular window, $x(n + m)$ is replaced by $x(n + m - N)$ when $(n + m)$ exceeds N. In the present case, the correlation between $x(n + m + N)$ and $x(n + m)$ is small and the averages of $x(n + m)x(n + m - N)$ are also small. As a result, the circular correlation problem does not affect the calculated results much when the auto-correlation is calculated by the FFT.

Let us check a case with large auto-correlations for relatively large m. The impulse response $h(n)$ is made of three decaying sine waves with a time constant of 500 sample time intervals. The input is the same random sequence $r(n)$ as in Fig. 2.14. The output signal $x(n)$, the convolution of $r(n)$ and $h(n)$, is used as the signal for the analysis. Those are shown in the top three charts of Fig. 2.15.

The auto-correlation function calculated by method (0) takes on much larger values for large lags than the previous case. The power spectrum obtained by employing the FFT of the 512 point auto-correlation data is shown on the right

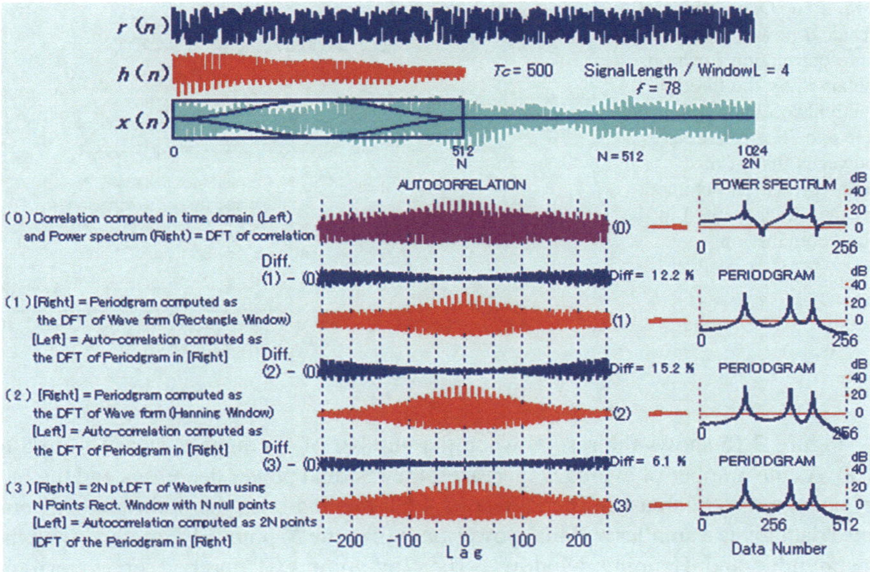

Fig. 2.15 Random input sequence $r(n)$ to a system with an impulse response $h(n)$ and the system output $x(n)$ (*top three charts*) and auto-correlation functions (*left*) and power spectra (*right*) calculated by the four methods from (0) to (3) (four *bottom charts*). Animation available in supplementary files under filename E9-15_AUTOCOR.exe

side. It has three peaks at the frequencies of the sine waves. The power spectrum has large values at frequencies above the highest peak frequency. This is because the auto-correlation function has large values at $m = \pm 256$. The FFT with the rectangular window of this auto-correlation function produces the spreading in the power spectrum.

In methods (1)–(3), the periodic spectra are calculated first and then the auto-correlation functions are obtained by IFFT. The residual power ratio of method (3) is the smallest, but still is over 6 %.

The power spectra of method (0) and (3) look very different, but the residual power ratio is relatively small. This is because the power spectrum is shown in decibels. Even if two values in the low levels have large difference in decibels, the residual power ratio is small if two values around the peaks are small.

2.10 Discussions of Stability and Estimation Precision

First, we will investigate the dependence of the residual power ratio, which was defined in the previous section, on the number of averages.

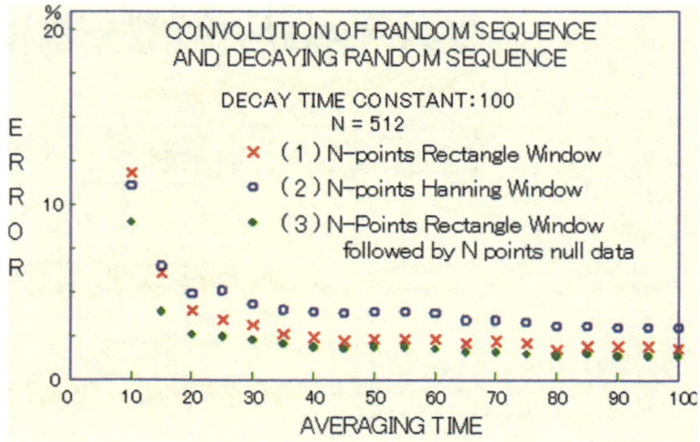

Fig. 2.16 Dependence of residual power ratios of the auto-correlation functions obtained by the three FFT methods on the number of average. The reference power is that of the auto-correlation obtained by the direct method in the time domain

Figure 2.16 shows the results when the number of averages is varied from 5 to 100. As the number of averages increases, the residual power decreases and the rate of change also decreases. The 2 N-point FFT method with the additional N-point zero data gives a smaller residual power ratio than the N-point FFT method with the rectangular and Hanning windows. The 2 N-point FFT method gives residual power ratios lower than 3 % if the number of averages is more than 20.

Shown above is the case with $N = 512$ and the time constant is equal to 100. Let us see how the results change if the time constant is set at different values. The random sequence $x(n)$ is the same as in Fig. 2.15, which is the convolution of the impulse response $h(n)$ comprising three decaying sine waves and the random sequence $r(n)$.

Ratios of the sine wave frequencies are 1:1.9:2.7, and the number of cycles of each wave in the window takes on a non-integer value. The same time constant for these three waves is varied from 10 to 100 in 10 steps. Since the length of $h(n)$ is known to have a very small effect on the results, it is set equal to 5,120. One random sequence generated by a computer is used throughout this example since it also has a very small effect on the result.

Residual power ratios of auto-correlations calculated by the three FFT methods to auto-correlations calculated by the direct time domain method are shown in Fig. 2.17. The results for (1) N-point FFT with the rectangular window, (2) N-point FFT with the Hanning window, and (3) 2 N-point FFT with additional N-point zero data are shown by ×, ○, and ●, respectively. Residual power ratios for one random sequence $x(n)$ were calculated as functions of the decay constant. The eight curves for each case shown in the figure were obtained for eight different random sequences.

The results show the following. All three methods have small residual powers if the time constant is below 50 (less than 1/10 of the window length). This is a proof that the effect of the circular correlation (replacement of $x(n + m)$ by $x(n + m - N)$ when $n + m$ exceeds N) is small because of the small correlations for m (lag) $> N$ if the time constant is much shorter than the window length.

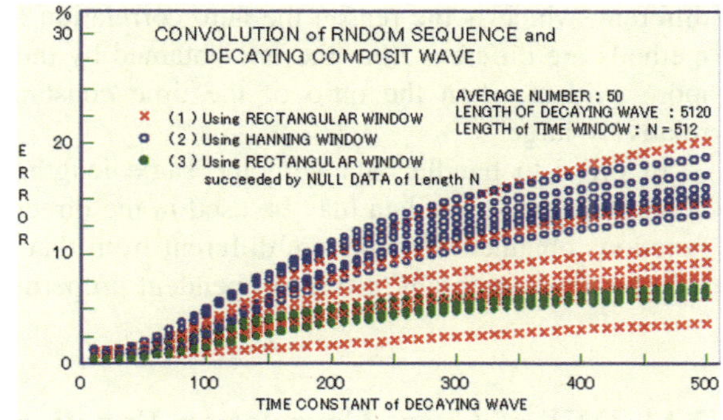

Fig. 2.17 Dependence of residual power ratios of auto-correlations obtained by the three FFT methods on the time constant

For large time constants, the correlations for large lags become large and the effect of the circular correlation becomes non-negligible. In this case, the differences of the residual powers among the three methods become visible.

In the first method, using the *N*-point FFT with the rectangular window, the residual power ratio changes drastically depending on the frequencies of the sine waves as shown by × in Fig. 2.17. If the frequencies of the sine waves are kept constant, the residual power ratio changes smoothly as the time constant increases. However, in this method, the starting and ending values of $x(n)$ of each window have large effects on this curve and, therefore, the curves for different sequences take quite different paths.

The second method, using the Hanning window to reduce the starting and ending values near the edges of the window, may be a good choice for the reduction of the unstable results of the first method. The curves shown by ○ are the results of the second method. The variation of the results is reduced but the magnitudes of the residual powers are not small compared to other two methods. This is because the auto-correlations for larger lags sharply reduce due to the window, compared to those of the direct time domain method.

The above mentioned variations are removed in the third method since $x(n+m)$ becomes always zero when $(n+m)$ exceeds N. The value of auto-correlation reduces linearly from $m=0$ to $m=\pm N$. Accordingly, the residual power ratio gets as large as approximately 1/12 even for large time constants (see Appendix 2D). Curves indicated by ● in the figure show this. Small variations still exist since the random sequences are different for different calculations.

The conclusion from the above discussion is that the IDFT of the averaged power spectrum obtained by the 2 *N*-point FFT method with the additional *N*-point zero data gives a good and stable estimation of the true auto-correlation function. The averaged power spectrum obtained by this method is a good estimation of the true power spectrum but still some amount of bias error cannot be avoided.

The time domain calculation of the auto-correlation function follows the formula Eq. (2.12). On the other hand, the FFT methods use the finite sequence lengths, N or $2N$. Data used in the direct method and the FFT methods are different, which is the reason the auto-correlation functions obtained by the FFT methods are different from the one obtained by the direct method. This becomes more evident when the ratio of the time constant to the window length gets relatively large.

In order to handle data with the same length, finite sequences with length N with external zero data may be used in the direct method. The auto-correlation functions obtained this way are different from that of the long sequence but they may be more preferable if time dependent properties of data are investigated.

2.11 DFT of Cross-Correlation Functions

Calculations of the cross-correlation functions and the related functions are similar to those of the auto-correlation functions. So, let us start from the calculation of the DFT of the cross-correlation function defined as Eq. (2.20). Since the cross-correlation function is obtained from N-point sequences by the same procedure as the auto-correlation function, the DFT of the cross-correlation function is calculated similarly as

$$R_{xy}(k) = \mathrm{DFT}[r_{xy}(m)] = \frac{1}{N\sigma_x\sigma_y}\sum_{m=0}^{N-1}\sum_{n=0}^{N-1} x^*(n)y(n+m)\exp(-j2\pi\frac{mk}{N})$$

The above equation is rewritten as (see Appendix 2E)

$$R_{xy}(k) = \frac{1}{N\sigma_x\sigma_y}X^*(k)Y(k) \tag{2.27}$$

or

$$X^*(k)Y(k) = N\sigma_x\sigma_y R_{xy}(k) \tag{2.28}$$

Equation (2.27) means that the DFT of the cross-correlation function between $x(n)$ and $y(n)$ is equal to $X^*(k)Y(k)$, the complex conjugate of $X(k)$ (DFT of $x(n)$) multiplied by $Y(k)$ (DFT of $y(n)$) and divided by $N\sigma_x\sigma_y$. The product of $X*(k)$ and $Y(k)$ is the cross-power spectrum, which will be discussed in Chap. 3.

The IDFT of Eq. (2.27) gives Eq. (2.29) which enables the calculation of the cross-correlation function from the cross-power spectrum.

$$r_{xy}(m) = \frac{1}{N\sigma_x\sigma_y}\mathrm{IDFT}[X^*(k)Y(k)] \tag{2.29}$$

The cross-power spectrum in this equation is the one calculated by the DFT of the cross-correlation function, therefore, Eq. (2.29) should not be interpreted as an equation that gives the cross-correlation function from the cross-power spectrum.

It must be considered carefully whether the cross-power spectrum obtained by the DFTs of sequences can be applied to the right-hand side of Eq. (2.29) to give the cross-correlation function.

The cross-power spectrum $X^*(k)Y(k)$ calculated by N-point DFTs is complex and its real and imaginary parts take both positive and negative values. The values given by averaging many cross-power spectra of two uncorrelated sequences will approach zero as the number of averages increases. If we consider only this aspect, it seems that the cross-power spectrum obtained by employing many averages can be used for the right-hand side of Eq. (2.29). This is different from the case of the auto-power spectrum.

The circular correlation problem of the DFT also exists in the cross-power spectrum. If there is a way of avoiding or reducing this, it may be possible to replace $X^*(k)Y(k)$ by using many averages of the cross-power spectrum obtained by selecting continuous N-point sequences from $x(n)$ and y(n).

$$r_{xy}(m) = \frac{1}{N\sigma_x\sigma_y}\mathrm{IDFT}\left[\overline{X^*(k)Y(k)}\right] \tag{2.30}$$

The bar above $X^*(k)Y(k)$ indicates that many averages have been taken. By employing this equation it becomes possible to calculate the cross-correlation function by using the N-point FFTs and IFFTs of the two sequences with much fewer computations.

If $x(t)$ and $y(t)$ are continuous time functions, the relation shown below exists between the cross-correlation and the cross-power spectrum (see Appendix 2F).

$$R_{xy}(\tau) = \frac{1}{\sigma_x\sigma_y}X^*(f)Y(f) \tag{2.31}$$

If a random sequence $x(n)$ goes through a transfer system and results in $y(n)$ as its output, the cross-correlation between $x(n)$ and $y(n)$ is the impulse response $h(n)$ of the system. A simulation example is shown in Fig. 2.4. A random sequence $x(n)$ is input to the same transfer system with the impulse response $h(n)$, and the cross-correlation function between the input $x(n)$ and the output $y(n)$ is calculated. The results are shown in Fig. 2.18. The top three charts are for $x(n)$, $h(n)$ and $y(n)$, respectively. The left side of Fig. 2.18(0) shows the cross-correlation function directly calculated in the time domain after 1,000 averages. In this simulation, the spacing between adjacent sets of sample sequences are varied between 7 and 19. The DFT of the cross-correlation function is shown by the right side of Fig. 2.18(0). This corresponds to Eq. (2.27).

The cross-power spectrum is a function made from a real even function and an imaginary odd function. They are shown separately in the right side of the figure only in the positive frequency region. Residual power ratio is also shown at the right corner of each correlation function (impulse responses). The residual power ratio is the sum of the squares of differences between the sample values of the estimated and the true impulse responses to the sum of the squares of the sample

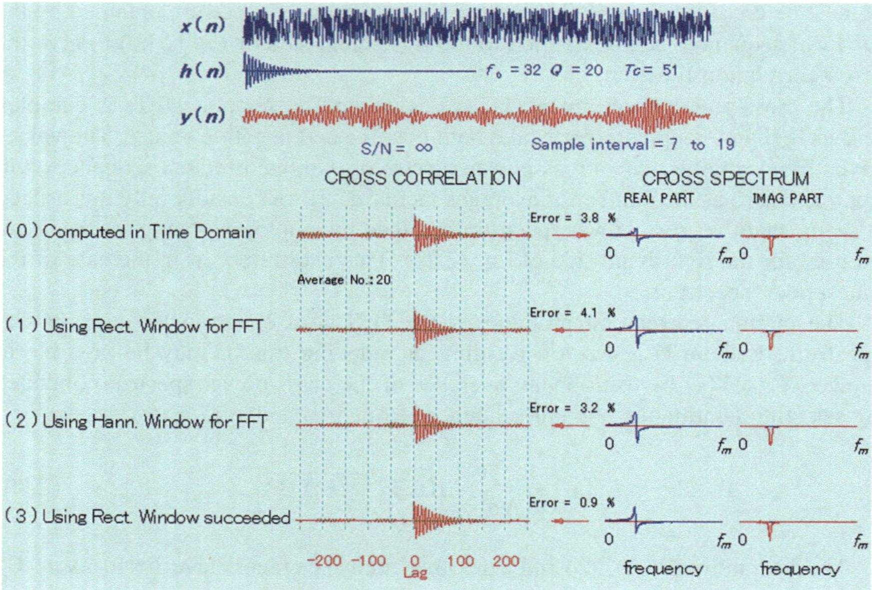

Fig. 2.18 Random input sequence $x(n)$ and its output $y(n)$ through a system with an impulse response $h(n)$ (*top three charts*) and the cross-correlation functions and corresponding cross-power spectra obtained by four methods (*bottom four columns*). Animation available in supplementary files under filename E9-18_CrosCorFT.exe

values of the true impulse response in the positive time region. Fig. 2.18(0) shows the case when the cross-correlation function is calculated in the time domain. The residual power ratio is 3.9 %.

The next three results are obtained using Eq. (2.30) in a similar manner to those discussed in Sects. 2.9 and 2.10. The residual power ratios of Fig. 2.18(1–3) are 4.1, 3.2 and 0.9 %, respectively. The last method (3), that avoids the circular correlation problem, gives much smaller errors than the other methods.

The cross-correlation between the input and output becomes the impulse response of the system only when the auto-correlation function of the input sequence is equal to one at m (lag) = 0 and 0 elsewhere. If not, the impulse response cannot be estimated from the cross-correlation function.

Figure 2.19 shows one example of this case. The input sequence of this example is a 4-point moving average of a random sequence. Only the input is changed from Fig. 2.18. It can be seen that the residual power ratios are much larger than the previous case. This difference is caused by the averaging of the input sequence only over the 4-point width.

It was shown in Fig. 2.11 that the auto-correlation of a moving-averaged random sequence becomes an isosceles triangle with a base that has twice the width of

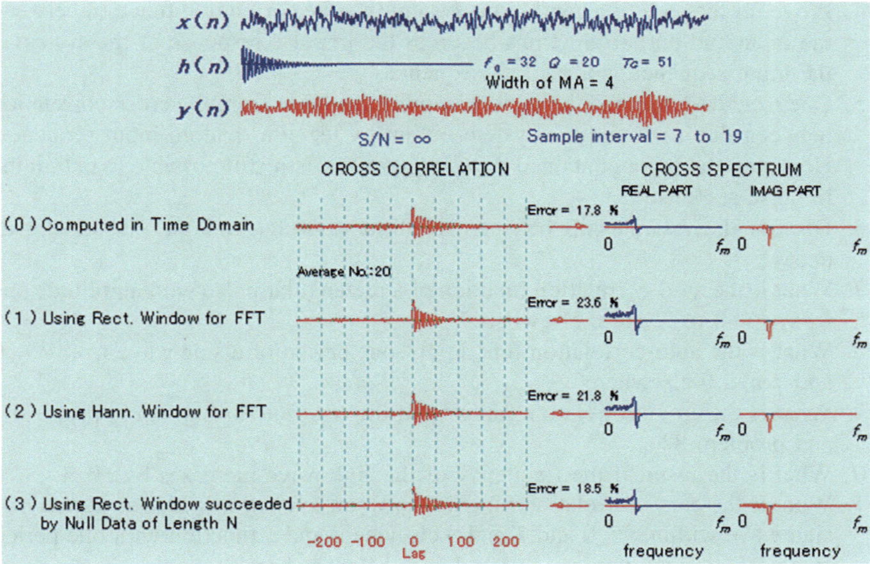

Fig. 2.19 Random input sequence $x(n)$ with 4-point moving average and its output $y(n)$ through a system with an impulse response $h(n)$ (*top* three charts) and the cross-correlation functions and corresponding cross-power spectra obtained by four methods (*bottom* four columns). Animation available in supplementary files under filename E9-19_CrosCorFT.exe

the averaging. In order for the cross-correlation function to be the impulse response, the width of the isosceles triangle must be infinitely narrow. In other words, the input must be perfectly uncorrelated (auto-correlation must be zero except at $m = 0$). This condition cannot always be expected. If the estimation of the impulse response is an objective, we should consider other methods. That will be discussed in Chap. 3.

In the programs associated with the figures, the student can choose various conditions and numbers. By trying various cases, the student will learn more that is not discussed in this chapter.

2.12 Exercise

1. What is the standard deviation of a sine function with amplitude 1?
2. Why is Eq. (2.4) equal to two when x(n) and y(n) are completely uncorrelated?
3. Why is Eq. (2.4) equal to four when x(n) and y(n) have the same magnitudes but their signs are opposite?

4. Prove for the case of a continuous system that the correlation function between the input and output functions becomes the impulse response of the system if the input sequence is a random sequence.
5. One tried to determine the system impulse response from the cross-correlation between the input and the system output by using a random input sequence. However, the input contained a DC component. Is it still possible to obtain the impulse response?
6. Discuss the relationship between the time constant and the period of half decay.
7. What is the auto-correlation function of a rectangular pulse with amplitude one from $t = 0$ to T and 0 elsewhere?
8. What is the auto-correlation function of one period of a sine wave from $t = 0$ and T and 0 elsewhere?
9. What is the cross-correlation function of the two functions given in problem 7 and problem 8?
10. What is the result if the frequency of the sine wave increases by 50 %?
11. What is the cross-correlation function between a function with one period of a sine wave within $t = 0$ and T and 0 elsewhere and a function with one period of a cosine wave within $t = 0$ and T and 0 elsewhere?
12. What is the result if the numbers of waves in the period from 0 to T are multiplied by 1.5?
13. What is the result if the number of sine wave in the period from 0 to T is multiplied by 1.5 and the number of the cosine wave in the same period is multiplied by 2.3?
14. When the cross-correlation is calculated using Eq. (2.8), there is a risky case if P is constant. What is it?

Chapter 3
Cross-Spectrum Method

The main subject of Chap. 1 was that the convolution of the system input signal and the system impulse response is the system output signal. It was shown that the Fourier transform of the convolution is given by the product of the Fourier transforms of the input signal and the impulse response. The main topic of this chapter is to obtain the system transfer function, which is the Fourier transform of the system impulse response, from the input and output signals utilizing these properties.

The transfer function or the impulse response of a system is obtained as a solution of a differential equation that describes the behavior of the system. However, such cases are very rare in practical situations. The estimation of the transfer function or the impulse response from the input and output signals is very important.

3.1 System and Its Input and Output

As a starting point, we must make clear the relation between the input and output of a transfer system, which will be shown here as a review.

Figure 3.1 shows a transfer system and its input and output. The signal $x(t)$ is input to the system and its response of the system $v(t)$ is the output. Both are functions of time t, i.e., they are waveforms. What is possibly observed is not $v(t)$ but $v(t)$ with an additional noise $n(t)$, which is denoted by $y(t)$. The Fourier transforms of the time functions are denoted by the corresponding capital letters as functions of frequency f.

The transfer system gives the output $y(t)$ by convolving the input $x(t)$ with the impulse response $h(t)$. The impulse response represents the performance of the transfer system. Its Fourier transform is the *transfer function* $H(f)$, which is also referred to as the *frequency response function*. The following relations hold:

K. Kido, *Digital Fourier Analysis: Advanced Techniques*,
DOI: 10.1007/978-1-4939-1127-1_3,

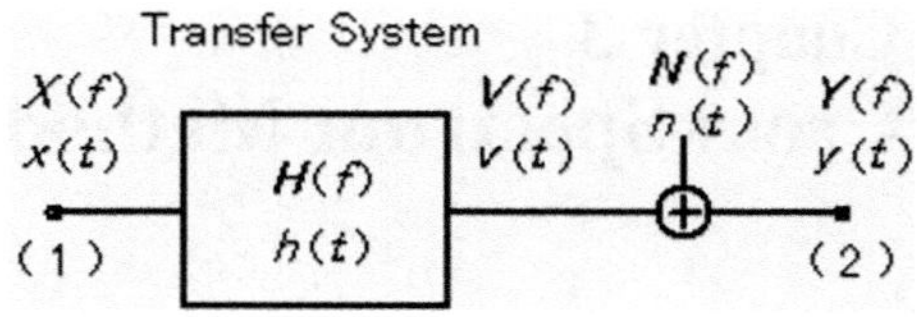

Fig. 3.1 Transfer system and its input and output

$$H(f) = \mathrm{FT}[h(t)]. \tag{3.1}$$

$$h(t) = \mathrm{IFT}[H(f)] \tag{3.2}$$

Using the transfer function and the impulse response, the relation between the input and the output is represented in the time and frequency domains, respectively, by:

$$v(t) = x(t) * h(t) \tag{3.3}$$

$$V(f) = X(f)H(f) \tag{3.4}$$

Since we assume that the output $y(t)$ is contaminated by noise $n(t)$, the relation between the input and the observed output are given by:

$$y(t) = x(t) * h(t) + n(t) \tag{3.5}$$

$$Y(f) = X(f)H(f) + N(f) \tag{3.6}$$

In this chapter, we will discuss the principle and related problems of the cross-spectrum method as a technique to estimate the impulse response $h(t)$ or the transfer function $H(f)$ from the input $x(t)$ and the output $y(t)$.

3.2 Principle of the Cross-Spectrum Method

Let us review the power spectrum that has been discussed many times from Chap. 2 in volume 1. The Wiener-Khinchine's theorem stating that the power spectrum is the Fourier transform of the auto-correlation was introduced in Chap. 2. The power spectrum is the sum of squares of the real and imaginary parts. The power spectrum $Wxx(f)$ of $x(t)$ is given by

$$W_{XX}(f) = X^*(f) \cdot X(f) \tag{3.7}$$

where $X(f)$ is the spectrum of $x(t)$ and $X^*(f)$ is the complex conjugate of $X(f)$.

The cross-spectrum of $x(t)$ and $y(t)$ is defined as the multiple of $X^*(f)$ and $Y(f)$.

$$W_{XY}(f) = X^*(f) \cdot Y(f) \tag{3.8}$$

The power spectrum given by Eq. (3.7) takes nonnegative real values, but the cross-spectrum given by Eq. (3.8) is generally complex.

Multiplying $X^*(f)$ on both sides of Eq. (3.6) gives

$$X^*(f) \cdot Y(f) = X^*(f) \cdot X(f)H(f) + X^*(f) \cdot N(f)$$

Rewriting this in the forms of Eqs. (3.7) and (3.8), we have

$$W_{XY}(f) = W_{XX}(f)H(f) + W_{XN}(f) \tag{3.9}$$

where $Wxn(f)$ is the cross-spectrum of input $x(n)$ and the external noise $n(t)$.

If $Wxn(f) = 0$ in Eq. (3.9), the transfer function $H(f)$ is given by dividing the cross-spectrum $Wxy(f)$ by input power spectrum $Wxx(f)$. The division by $Wxx(f)$, which is real, is simple. However, $Wxx(f)$ should not be zero at any frequency since we cannot divide by zero. The physical meaning of this is that there is no way to know about the system performance at frequencies with no input.

Our question is whether $Wxn(f)$ becomes zero or not. If the external noise is independent of the input, their cross-correlation is zero and therefore the cross-spectrum is also zero (see Eq. (3.27)). As a result, $Wxn(f) = 0$ is satisfied and Eq. (3.9) becomes:

$$W_{XY}(f) = W_{XX}(f)H(f) \tag{3.10}$$

or

$$H(f) = \frac{W_{XY}(f)}{W_{XX}(f)} \tag{3.11}$$

Equation (3.11) shows that the transfer function is obtained by dividing the cross-spectrum between the input and output by the power spectrum of the input. The cross-spectrum and the power spectrum are obtained by the Fourier transforms of the cross-correlation and the autocorrelations, respectively. The necessary condition is that the power spectrum of input is not zero at any frequency and the external noise is independent of the input signal.

Once the transfer function is obtained, the impulse response of the system is calculated by its inverse Fourier transform.

$$h(t) = \mathrm{IFT}[H(f)] \tag{3.12}$$

As described above, if the input and the output signals are observed, the system transfer function is obtained and the impulse response is also obtained by the inverse Fourier transform. The equations given above are those of the continuous domain and they are not convenient for numerical calculations.

One thing that worries us is that Eq. (3.11) has the power spectrum of the input as the denominator. It was so often mentioned since Chap. 6 in Volume 1 that the average of the periodogram does not give the power spectrum. If we must take the roundabout way of obtaining the auto-correlation function first and then apply the Fourier transform to it, the numerical operation load becomes much heavier.

The discussion in Chap. 2 indicated that, in some conditions, many times of averaging of periodogram of extracted finite sequences can be used as an approximation of the power spectrum $Wxx(f)$. This should be discussed more in detail.

Actually, there is another way of thinking, which will be introduced in Sect. 3.3.

3.3 Estimation of Transfer Functions

In the previous section, it was found that cross-spectrum between the input and the output signals divided by the input power spectrum is the transfer function of the system (Eq. (3.11)). However, this equation is for a continuous system and we need equations that can be applied to discrete input and output sample sequences.

We need to go back to Eq. (1.1), which gives the convolution between the discrete input signal $x(n)$ and discrete impulse response $h(n)$. Since the system in Fig. 3.1 contains the external noise $n(n)$ in its output $y(n)$, it is given as

$$y(n) = X(n) * h(n) + n(n) \tag{3.13}$$

An N-point sequence is chosen from $y(n)$ and DFT is applied to this sequence. In fact, there is some problem in this step, but we will consider it later. The DFT of Eq. (3.13) is given by

$$Y(k) = X(k)H(k) + N(k)$$

By multiplying $X^*(k)$ on both sides, we have

$$X^*(k)Y(k) = X^*(k)X(k)H(k) + X^*(k)N(k) \tag{3.14}$$

The left-hand side of Eq. (3.14) is the cross-spectrum between $x(n)$ and $y(n)$. The first term of the right-hand side is the product of the periodogram and the transfer function, and the second term is the cross-spectrum between $x(n)$ and $n(n)$. Following the notation in Chap. 6 in volume 1, Eq. (3.14) is represented as

$$P_{XY}(k) = P_{XX}(k)H(k) + P_{XN}(k) \tag{3.15}$$

The term $Wxn(f)$ in the continuous system is a cross-spectrum, which is zero if $x(t)$ and $n(t)$ are un-correlated. But $P_{XN}(k)$ is a cross-spectrum, which is calculated from finite sequences, and does not necessarily become zero. If $P_{XN}(k) = 0$ does not hold, the transfer function cannot be obtained.

If an average of the cross-spectrum is calculated by taking N-point sequences of $x(n)$ and $y(n)$ with different starting times, the term $P_{XN}(k)$ can be made equal to zero. By denoting operation of many times of averaging by the upper bar, Eq. (3.15) can be represented as

$$\overline{X^*(k)Y(k)} = \overline{X^*(k)X(k)}H(k) + \overline{X^*(k)N(k)} \tag{3.16}$$

Following the notation of Eqs. (3.15) and (3.16) can be written as

$$\overline{P_{XY}(k)} = \overline{P_{XX}(k)}H(k) + \overline{P_{XN}(k)} \tag{3.17}$$

The cross-spectrum of $x(n)$ and $n(n)$ is a complex number, but since it varies randomly with different sequences, the real and imaginary parts of $P_{XN}(k)$ also change randomly, and after many times of averaging both of them approach zero. If the average of $P_{XN}(k)$ becomes zero, the remaining process is simple. The transfer function is calculated by

$$H(k) = \frac{\overline{X^*(k)Y(k)}}{\overline{X^*(k)X(k)}} = \frac{\overline{P_{XY}(k)}}{\overline{P_{XX}(k)}} \tag{3.18}$$

Equation (3.18) has the same form as Eq. (3.11), but the numerator and the denominator are the averages of the cross-spectrum and the periodogram, respectively. Equation (3.18) is a practicable equation which is easily performed numerically.

The impulse response is obtained by the IDFT.

$$h(n) = \mathrm{IDFT}\{H(k)\} \tag{3.19}$$

Equation (3.19) shows that the transfer function is obtained by dividing the average of the cross-spectrum between the input and output by the average of the periodogram of the input. In the derivation of this process, it was assumed that the external noise $n(n)$ is random, but there is no restriction that the input sequence $x(n)$ must also be random. It can be said that the transfer function is obtained from Eq. (3.18) as far as the periodogram of the input $x(n)$ does not take zeros at any frequency.

However, we still have to worry about the circularity problem of DFT. Even if Eq. (1.16) says that the DFT of the convolution of the input sequence and the impulse response gives the discrete Fourier transform of the output, the IDFT of the product of the N-point DFTs of the input sequence and the impulse response does not give the correct output sequence as was discussed in Chap. 1. Keeping this in mind, we must consider that there might be a problem in calculating the impulse response from the IDFT of $H(k)$ given by Eq. (3.18).

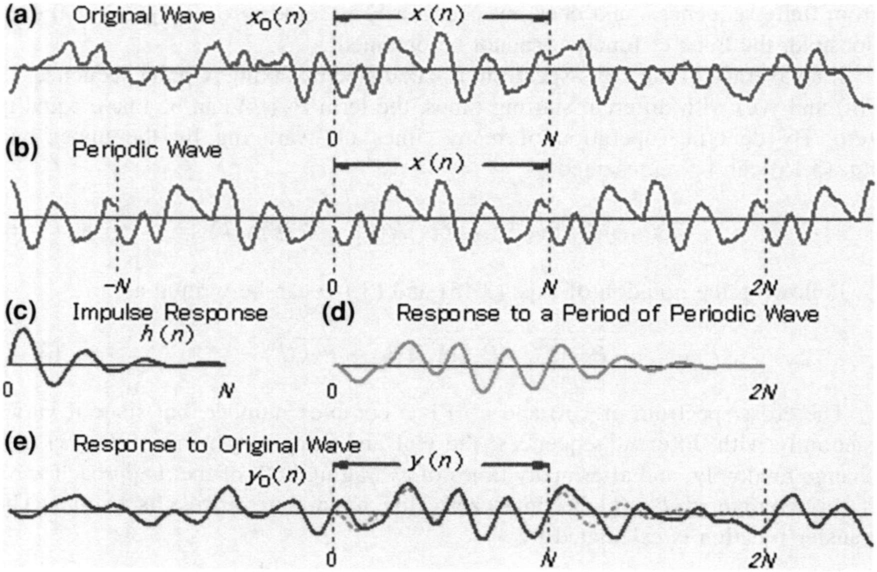

Fig. 3.2 Relation of signals when a long sequence is input to the system and finite (N-point) DFTs are used for the transfer function estimation. **a** Long original sequence $x_0(n)$, **b** Periodic sequence made of an N-point sequence $x(n)$ extracted from (**a**, **c** Impulse response $h(n)$, **d** Convolution of $x(n)$ and $h(n)$, **e** $y_0(n)$, the convolution of $x_0(n)$ and $h(n)$

Let us consider the relation between the input and the output using Fig. 3.2. Figure 3.2a shows a long input sequence $x_0(n)$. A finite N-point sequence $x(n)$ is chosen from this sequence. The periodogram $P_{xx}(k)$ gives squares of absolute value of each Fourier coefficient of an infinite periodic sequence made of $x(n)$ as shown by Fig. 3.2b.

The output $y_0(n)$ when the input $x_0(n)$ is given to the system with the impulse response $h(n)$ (Fig. 3.2c) is a long sequence (Fig. 3.2e) given by the convolution $x_0(n)*h(n)$.

In Eq. (3.18), $X(k)$ and $Y(k)$ are N-point DFTs of $x(n)$ (Fig. 3.2b) and $y(n)$ cut out from $x_0(n)$ (Fig. 3.2a) and $y_0(n)$ (Fig. 3.2e), respectively. We should note that $x(n)$ and $y(n)$ are, respectively, different from $x_0(n)$ and $y_0(n)$ where $y_0(n)$ is the convolution between $x_0(n)$ and $h(n)$. Unfortunately, $y(n)$ is not the exact convolution between $x(n)$ and $h(n)$, therefore, the exact $H(k)$ can not be obtained by Eq. (3.18) using $x(n)$ and $y(n)$.

Equation (3.16) is given as many times of averaging of $X^*(k)Y(k)$ while pairs of N-point sequences $x(n)$ and $y(n)$ are chosen by shifting the starting point on the time axis. Equation (3.18) shows that the transfer function $H(k)$ is given by dividing the average of $X^*(k)Y(k)$ by the average of the periodogram $X^*(k)X(k)$.

However, the convolution of the N-point input sequence $x(n)$ (Fig. 3.2a) and the impulse response $h(n)$ (Fig. 3.2c) exceeds the range from $n = 0$ to $N - 1$ as shown by Fig. 3.2c. The portion exceeds this range falls in the range from $n = N$ to $2N - 1$. But this portion of $y_0(n)$ is not used in Eq. (3.18). It only uses the portion of response $y_0(n)$ in the range from $n = 0$ to $N - 1$. Since there is some percentages are discarded in the calculation of each cross-spectrum, the average of $X^*(k)Y(k)$ cannot be a correct estimate of the cross-spectrum of $x_0(n)$ and y_0 (n).

The cause of the above-mentioned problem is that the impulse response is estimated by applying the DFT, which has the property of the circular convolution to the output sequence, which does not have this property. A method of avoiding this circularity problem must be devised.

3.4 Circular Convolution and Time Windows

As mentioned in the previous section, periodograms calculated from N-point sequences extracted from the input and output of a transfer system do not give correct transfer functions (and consequently impulse responses) since the circular convolution problem is introduced. This method will be referred to as method (1). It is necessary to propose second and third methods that will hopefully reduce the errors caused by the circularity problem.

The second method (method (2)) is to apply a time window, such as the Hanning window, which has tapering at both ends, to the extracted sequences. Let us discuss why this may work.

Most of the data that spills over into the range $N \le n < 2N$ is the response to the later part of the input $x(n)$. By applying the window, these parts of the input and output data are suppressed. At the same time, the early part of the output $y(n)$, which contains a response to the input sequence preceding the sequence $x(n)$, is also suppressed. In this way, the spillover problem will be reduced by tapering the input and output sequences. If the impulse response is of short duration, this method should be effective. It is hoped that the error would, at least, be reduced even if the problem is not completely solved.

The third method (method (3)) is easily derived considering the following procedure. If the length of the impulse response of the transfer system is within, say, N, the response to an input sequence with length N will be within $2N$. Therefore, if the first N-point sequence is chosen as the input and the $2N$-point sequence as the output, all responses to the input are included in the response, and nothing is lost. Further, if an N-point zero data is added to the input sequence, the data that are circulated to the front end of the output becomes zero and the response to them is also zero. In this way, the error caused by this circular convolution problem is removed.

In the above method, there are still two sources of error. In the latter half of the $2N$-point response $y(n)$, a response to the input sequence from $n = N$ to $2N - 1$ is included. However, since this is not the response to the $2N$-point input sequence $x(n)$, it is uncorrelated noise. Also, in the first half of the $2N$-point response $y(n)$, a response to the preceding input sequence is included, which is also uncorrelated noise.

In the third method, there are $2N$-point data in the same frequency range because the $2N$-point DFT is used. Since the length of the time sequence is doubled, the sampling period in the frequency domain (frequency resolution) becomes 1/2 and the maximum frequency given by multiplying the frequency resolution by $2N$ is unchanged. The sampling period of the impulse response calculated by the $2N$-point IDFT is the same as that of the original time sequences.

The reader may have noticed that method (3) is the same as the one introduced in Chap. 1, which gives a correct response of the system; also that the three methods introduced here are the same as the three methods described in Chap. 2. Discussion of the results obtained by these three methods will now be undertaken.

Figure 3.3 shows the input waveform, the impulse response, and output waveform of the transfer system model in the upper three rows, respectively. Below these rows, transfer functions calculated by Eq. (3.18) and its IDFTs using the three methods (1), (2), and (3), respectively, are shown.

The input sequence $x(n)$ is a random sequence generated by a computer. The impulse response is a random sequence with an exponential decay (which may not be realistic but theoretically possible). The output $y(n)$ is the convolution calculated from $x(n)$ and $h(n)$.

Many pairs of 256-point sequences of $x(n)$ and $y(n)$ both with common starting points are chosen from $x(n)$ and $y(n)$. Then $X^*(k)Y(k)$ and $X^*(k)\,X(k)$ are calculated for each pair and their averages are calculated. Finally, the transfer function is obtained using Eq. (3.18) and the impulse response is obtained by the IDFT. These are calculated using the three methods (1), (2) and (3).

Results obtained by method (1), i.e. using the 256 point rectangular window, are shown in Fig. 3.3 (1). The transfer function has both real and imaginary parts. The estimated impulse response is the 256-point IDFT of this transfer function. The imaginary part of the IDFT is zero as a matter of course. The residual power ratio, defined as the ratio of the sum of squares of the differences at 256 points between the estimated impulse response and $h(n)$ to the sum of squares of the impulse response $h(n)$ at 256 points, is 3.9 % for method (1).

Figure 3.3 (2) shows the transfer functions and impulse responses obtained by method (2), which applies the Hanning windows to the 256 input and output sequences. The residual power ratio is 6.3 %, which is larger than for method (1). It will be shown later that the error becomes smaller than method (1) as the time constant of the exponential decay of the impulse response becomes smaller.

Figure 3.3 (3) shows the results obtained by method (3). The estimated transfer function has 512-point real and imaginary parts. Since the actual frequency range of method (3) is the same as those of method (1) and (2), they are shown with the same length in Fig. 3.3. The 512-point IDFT of the transfer function gives an impulse response that is two times longer than those of method (1) and (2). However, since the original impulse response has 512 point length, the residual power ratio is calculated using the first 256-point data. It is 0.5 %, which is much smaller than those of method (1) and (2).

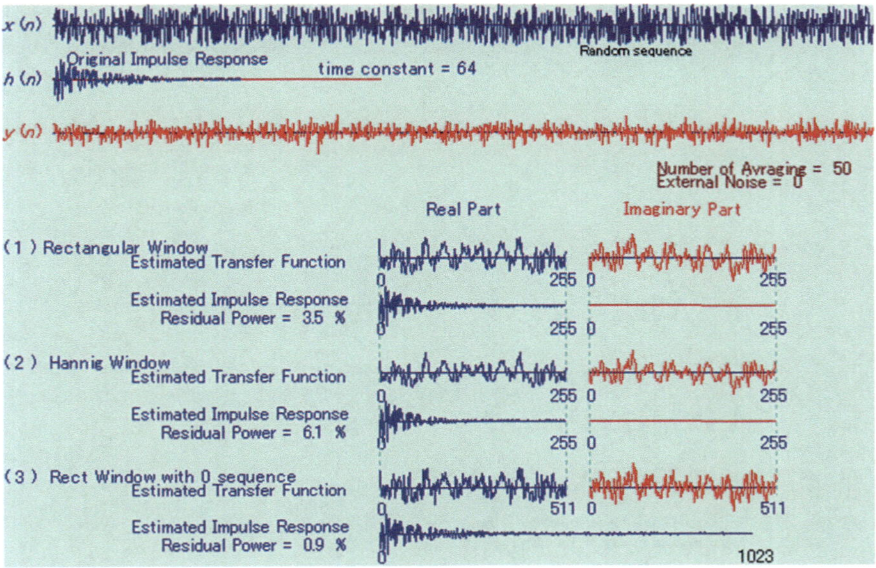

Fig. 3.3 Transfer function and an impulse response estimation when a random sequence is used as an input to a transfer system. Animation available in supplementary files under filename E10-03_Csp.exe

In the above example, the N-point input sequence (with additional N-point zero sequence) and $2N$-point output sequence are used, but this is not always necessary. The necessary condition is that the difference between the output sequence and the input sequence lengths should be larger than the length of the impulse response.

Method (2) produced the largest error. However, it produces a smaller residual power ratio than method (1) if the impulse response decays faster than the present case. This can be checked by running Program 10.3 with shorter time constants. The cause of the large error of method (2) in the present case is due to a large deformation of the impulse response, as will be discussed later.

Compared to the other two methods, method (3) is very effective. However, if the impulse response decays quickly, circular convolution is not the main cause of the error and method (3) is not necessarily the best method. This is because of the increased noise introduced by the use of data contained in the latter half of output $y(n)$.

The input sequence used in the example of Fig. 3.3 is random. This type of signal makes it possible to obtain accurate impulse responses. However, the randomness of the input signal is not a necessary condition. An impulse sequence of constant amplitude with randomly varying spacing can also be used as the input.

One example is shown in Fig. 3.4. The input sequence $x(n)$ used here has the properties that: (1) the amplitude is constant, (2) the spacing is randomly chosen

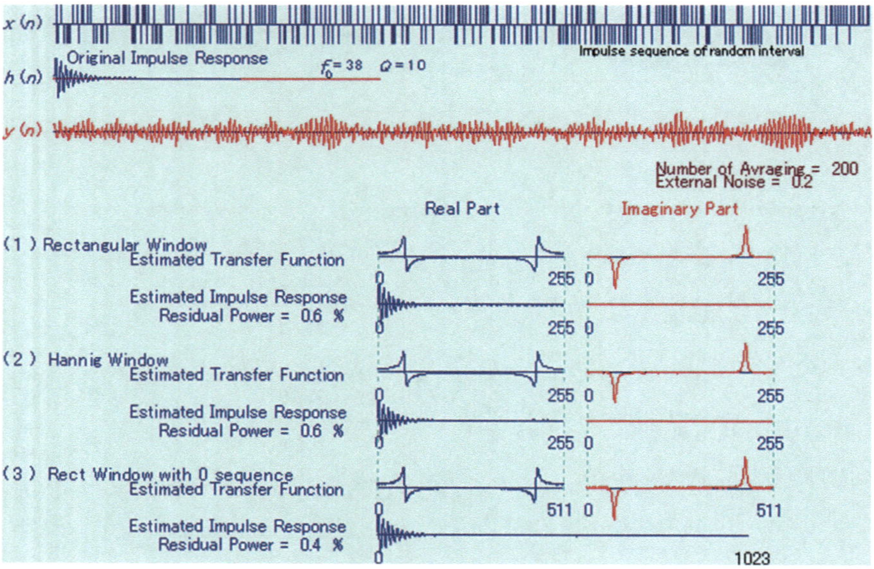

Fig. 3.4 Transfer function and impulse response estimation by the use of a random spacing constant amplitude alternating impulse sequences as the input signal to the transfer system. Animation available in supplementary files under filename E10-04_Csp.exe

from 1 to 7 sample lengths, (3) the sign of each pulse is randomly varied so that the long-term average is zero. The impulse response $h(n)$ is a decaying sine wave with $f_0 = 38$ and Q (quality factor) $= 10$. The time constant for this case is given by $NQ/(\pi f_0) = 0.038N$, which is roughly equal to 21 data intervals. This is approximately 1/3 of the previous case. External random noise with 1/5 the amplitude is also added in the present case.

The results in Fig. 3.4 were obtained by the same methods as in Fig. 3.3. However, since external noise is added, the number of averages is increased to 200.

The estimated transfer functions and impulse responses show that the system has one resonance with a quick decay. If the duration of the impulse response is sufficiently short relative to the window length, the effect of circular convolution becomes negligible and there is no need to take any action to avoid it. All three methods produce residual power ratios less than 1 %. If the time constant is larger than this case, method (3) is more advantageous, and if the time constant is shorter, method (2) is more advantageous.

Even for cases with large time constants (i.e., with small decay rates), method (3) explained in this section makes it possible to avoid the circular convolution problem of the DFTs in Eq. (3.18). However, one problem still remains.

3.5 Causality and Input Sequence

Let us assume a case where two signals with identical waveforms but with a time difference are input to a transfer system. If there is no way to distinguish which of the two responses is due to the earlier input, there may be a misinterpretation that the output has appeared before the input. If the input is not purely random and its auto-correlation takes large values not only at time 0, but also at other times, this kind of problem may occur. This is the present concern.

If this problem is considered from the standpoint of the principle of the cross-correlation method, it does not seem be a problem. The reason is simple.

If an input to a transfer system is an addition of one signal and an identical signal with time delay τ, the input signal is represented by

$$x_D(t) = x(t) + x(t - \tau)$$

and its Fourier transform (spectrum) is given by

$$X_D(f) = X(f)\{1 + \exp(-j2\pi f\tau)\}$$

In the derivation from Eqs. (3.7) to (3.11), the input signals can be replaced by $X_D(f)$ of the above equation and Eq. (3.11) is still valid.[1] Then, there is no need to worry about the causality problem. However, this is because Eqs. (3.7)–(3.11) deal with the Fourier transform in the infinite time domain. The following problem arises in finite time domain (DFT) analysis.

Figure 3.5 describes the situation. The N-point input and output sequences are sampled during the same time period. Figure 3.5a shows the case when the preceding signal $x(n)$ is in front of the window. The whole part of the m-sample delayed signal $x(n - m)$ is in the same window. Some part of the response to $x(n)$, shown by the hatched lines, is included in the window as shown in Fig. 3.5b. The response to $x(n - m)$ is shown by $r(n - m)$, which is identical to $r(n)$ but delayed by m-samples.

The signal $x(n)$ is not in the window, while the m-sample delayed signal $x(n - m)$ is in the window. If the signals included in the window are examined, the signal indicated by the hatched lines has the same shape as that of the response to $x(n - m)$. The only difference is that the hatched line part is leading the input $x(n - m)$. The real response to $x(n - m)$ is also included in the window, but of course it lags behind $x(n - m)$. The hatched line part is ahead of the input $x(n - m)$ and could be considered as a response before the input takes place, which means that the law of causality is broken. The impulse response calculated including this type of output will have a response, which seems to defy causality.

---

[1] The multiplication of $X(f)$ by $[1 + \exp(-j2\pi f\tau)]$ causes periodic zeros in the input. But this has nothing to do with causality.

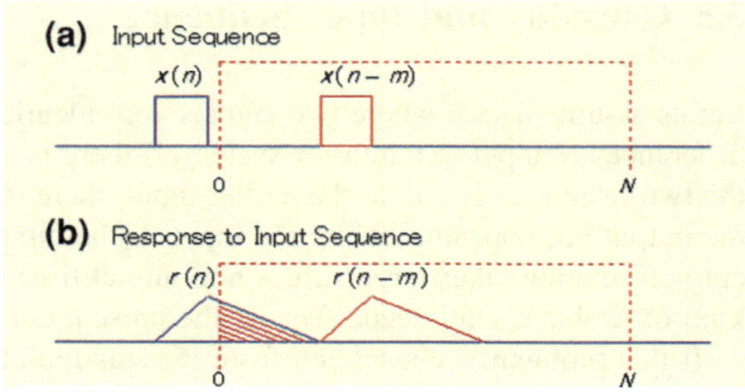

Fig. 3.5 Input and output signals near the time window for an input signal $x(n) + x(n-m)$

If both $x(n)$ and $x(n-m)$ are included in the input window and their responses are also included in the simultaneous output window, this type of problem will not occur.

When the input signal $x(n)$ is near the end of the input window and *some portion of* its response is outside of the output window, it is *discarded (interpreted as zero) in the impulse response estimation.* However, this does not violate causality.

Check the impulse response obtained by the cross-spectrum method using an input signal made up of a superposition of original and delayed sequences to see whether the result contains a component that violates causality.

Figure 3.6 shows the result. The input signal in this case is made of a 10-point moving average of a random sequence generated by a computer. Figure 3.6 (1) is the result obtained by method (1), which uses the rectangular window. The impulse response of Fig. 3.6 (1) contains a small ripple at the end. This part is not the end of the impulse response, but is the response occurring in the negative time domain. This appears to be in violation of causality. Method (2) using the Hanning window yields a smaller component in the negative time domain. This is because the component that is responsible for the break of causality is reduced by the window. The result of method (3) also has a small ripple similar to that of method (1). This component, which is removed if only the first half of the response is used, is generated for the same reason as in method (1). The appearance of components at the end of the estimated impulse response is frequently experienced when impulse responses are experimentally obtained.

As the above example shows, if the input signal is composed of correlated signals, the estimated impulse response obtained by the cross-spectrum method contains response components that seem to break the law of causality. When a signal generated by an actual machine is used as the source signal, it is common to see this kind of phenomenon. This is because the noise has components that are correlated with each other.

It has been found that the selection of the source signal is important in the estimation of the impulse response by the cross-spectrum method. If artificially generated signals can be used, those signals used in the examples of Figs. 3.3 and 3.4 should be adopted. Or the method that will be described in Sect. 3.6 should be used.

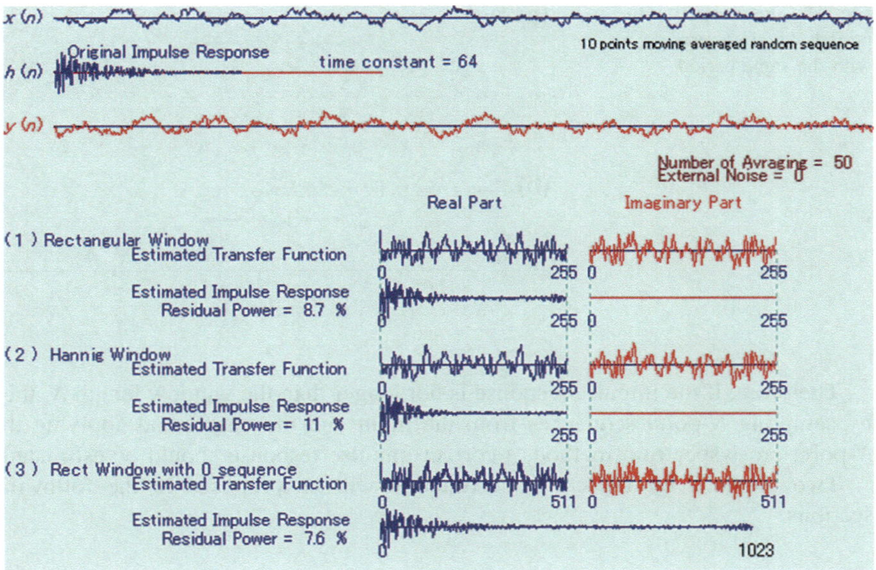

Fig. 3.6 Transfer function and impulse response estimation when a 10-point moving averaged random sequence is used as an input signal. *1* Rectangular window (method (1) in the text), *2* Hanning window (method (2) in the text), *3* Rectangular window with added zeros (method (3) in the text). Animation available in supplementary files under filename E10-06_Csp.exe

3.6 Utilization of Circular Convolution

In the previous discussions, it was found that there are several factors that cause errors in the impulse response estimation. They are (1) usage of finite length time windows, (2) circularity of the DFT, and (3) auto-correlation of the input signal. With regard to the second factor, a tactic that takes into account the circularity of the DFT itself can be implemented to reduce the estimation error.

Figure 3.7 a and b show an input signal with period N on the discrete time axis and its model output through a transfer system. If the input is a periodic signal with period N, there exist two identical N-point sequences before and after a specific time on the time axis. The output is also periodic with period N as long as the transfer system is time-invariant. In the figure, extracted N-point input and output sequences are denoted by $x(n)$ and $y(n)$.

The portion of the response caused by the later part of the input falls into the next window from N to $2N - 1$. The same portion of the response corresponding to the input in the window from $-N$ to -1 comes into the response window from 0 to $N - 1$, which seems as if it is a response to the input in the input window from 0 to $N - 1$.

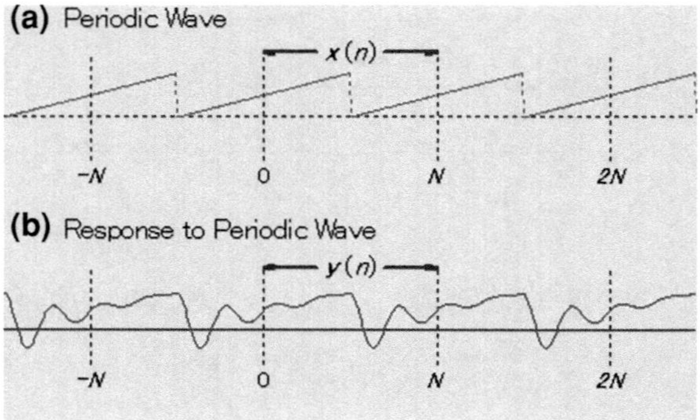

Fig. 3.7 Input signal with period N and its response with the same period

Therefore, if the impulse response is not longer than the window length N, then by sampling N-point sequences from the input and the output and applying the N-point cross-spectrum method, a correct impulse response should be estimated.

Two methods for the correct estimation will be proposed in the following sections.

3.6.1 Duplex Random Signal Source

The first method is to use a random sequence as the input signal. First, assuming that a random sequence is available, the input is the N-point sequence extracted from the original random sequence to the system from $n = -N$ and continue without a break from $n = 0$. The $2N$-point input sequence is called a *duplex random sequence*. The original point of time is set at the beginning of the second input. That is, the input signal $x_a(n)$ is from $n = -N$ to $n = N - 1$ and $x_a(n) = x_a(n - N)$, $0 \leqq n < N$. The N-point output of the system $y_a(n)$ is cut out from $n = 0$ to $n = N - 1$. $y_a(n)$ contains the response $y_{-a}(n)$ to the first input at $n < 0$, the response $y_{+a}(n)$ to $x_a(n)$ at $0 \leqq n < N$ and external noise $n(n)$. That is

$$y_a(n) = y_{-a}(n) + y_{+a}(n) + n(n)$$

where, $y_{-a}(n)$ is the same as the component produced by circular convolution.

Therefore, the exact impulse response should be obtained by the cross-spectrum method using Eqs. (3.18) and (3.19) so long as the length of the impulse response is shorter than N. Remember that both the input and the output sequence should be cut out from $n = 0$ to $n = N - 1$.

In the same way, other partial random sequences can be used $x_b(n)$ in $N < N_b \leq n \leq N_b + 2N - 1, x_c(n)$ in $N_b + 2N < N_c \leq n \leq N_c + 2N - 1, \ldots$ as shown in Fig. 3.8. Then, there are corresponding responses $y_m(n)$ to the input $x_m(n)$ at the output. From pairs of $x_m(n)$ and $y_m(n)$, the N-point periodogram and the cross-spectrum are calculated. Applying Eq. (3.18) to those averages, a correct transfer function and then an impulse response are obtained.

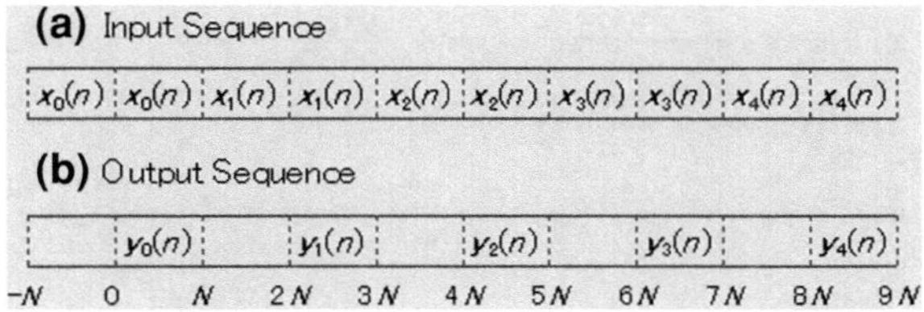

Fig. 3.8 Determination of input and output sequences for accurate impulse response estimation

One source of worry is whether there is any frequency at which the average of $P_{xx}(k)$ becomes zero. If $x_m(n)$ is a random sequence, it is impossible to know where and when that zero occurs. However, once a sequence is given, it is possible to produce another sequence that does not have a zero at the same frequency (see Appendix 3A). If a value at one sample point of the N-point sequence is changed, the frequency at which the zero occurs always changes. The new sequence can be used as the input to the next period.

If the input signal is made this way, the component to be generated by the circular convolution of DFT is given by the input at $n < 0$. Therefore, the averaging operation is not necessary unless external noise is mixed with the output signal.

Let us see if this is the case. In the impulse response estimation program of Figs. 3.3 or 3.6, if a periodic sequence with 256-point period is used as an input, a correct impulse response is obtained as shown in Fig. 3.9 with a single averaging, i.e., without averaging. Averaging is not necessary because the DFT of the source signal has no zero point and external noise is not included. Otherwise, averaging is necessary.

3.6.2 Periodic Swept-Sine Signal

A sinusoidal waveform whose frequency continuously changes is an example of a sequence with no "zeros" in a wide frequency range. If the signal has constant amplitude and constantly increasing frequency as a function of time, its spectrum is essentially constant up to the maximum frequency. If so, a sequence with its frequency starting from zero and reaching the maximum frequency within N samples can be used as an input signal not only for the measurement of the impulse response, but also for other applications.

Figure 3.10 shows the impulse response obtained by using the above-mentioned signal. Since the frequency changes continuously, this kind of signal is referred to as the *swept sine* wave. The power spectrum shown by Fig. 3.10a has an almost constant amplitude for over a wide frequency range. In this example, an external noise source with 20 % magnitude is added to the output, but after 200 averages, the residual power ratio between the estimated and true impulse responses is within 0.1 %.

The external noise ratio is 0.2 and the number of averages is 200. Animation available in supplementary files under filename E10-10_Csp.exe

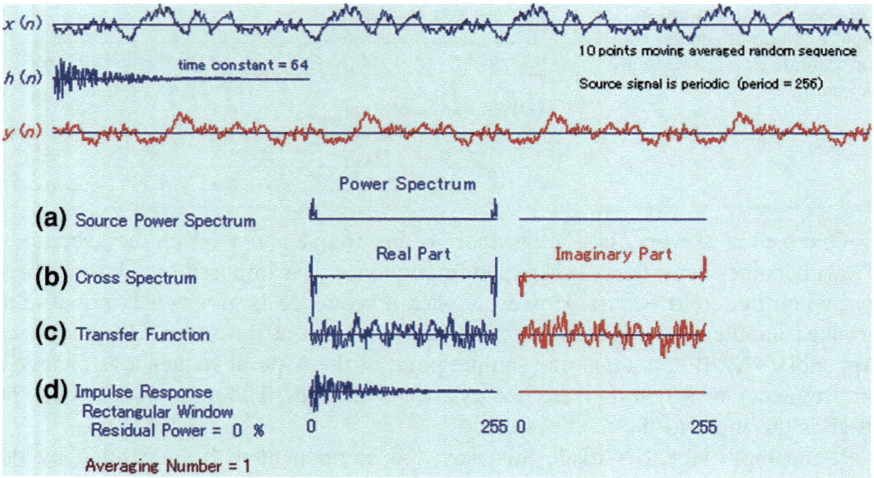

Fig. 3.9 Impulse response estimation by the use of a duplex random sequence with period of 256 with no external noise and no averaging. Animation available in supplementary files under filename E10-09_Csp.exe

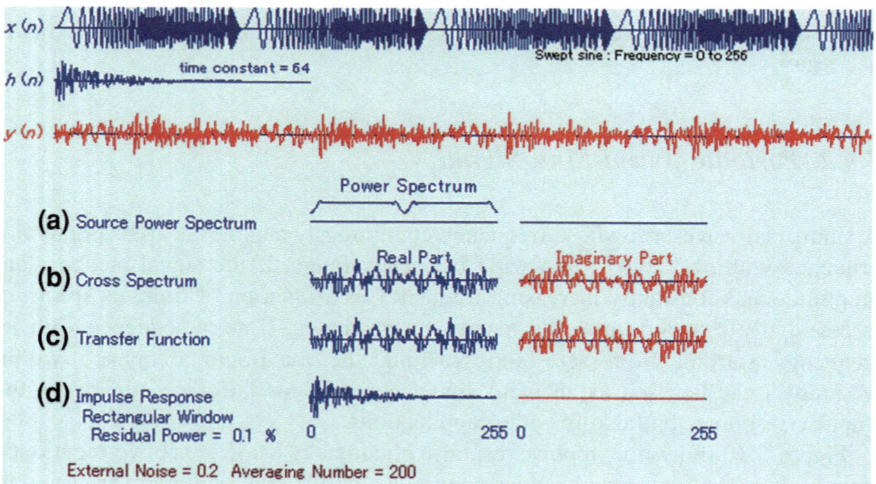

Fig. 3.10 Impulse response estimation by the use of swept sine wave as an input

The amplitude of the input sequence $x(n)$ shows some fluctuation as seen in Fig. 3.10. The reason is that there are only 256 sampling points in one period. However, the time waveform does not need to have a constant amplitude for the present purpose, and as far as the sampling theorem is satisfied, it is known from the discussion in Chap. 3 in Volume 1 that a continuous waveform can be reconstructed. The power spectrum computed from this sample sequence does not have a discontinuity, as demonstrated in Fig. 3.10a.

If each one-period sine wave from $x(n)$ in Fig. 3.10 is extracted from the waveform, it will be a sine wave with time delay proportional to the frequency. Since the delay on the time axis causes a phase delay that is proportional to the frequency, the waveform $x(n)$ is considered as one which has a phase delay that is proportional to the square of the frequency. If this waveform is fed into a filter which has a constant amplitude response and phase advance proportional to the square of the frequency, the output becomes an impulse. This means that the impulse and the swept sine wave are essentially the same. By adjusting the phase response, an impulse can be converted to a swept sine wave and a swept sine wave to an impulse. The swept sine wave is referred to as a *time stretched pulse (TSP)* in the sense that an impulse becomes a swept sine if the former is stretched in time. If a TSP is used as the input, the impulse response can be obtained by letting the output go through a filter with the inverse (sign-reversed) phase response of the input TSP. This gives an impulse response estimation method that does not depend on the FFT.

Since the swept sine wave can have high energy if the signal duration is made large, the use of the TSP can reduce the effect of external noise in the same way as a large amplitude impulse is used. This is effective when a large input signal to an electrical circuit cannot be handled because of the electrical circuit's nonlinearity. Thus, the swept sine is useful in a noisy measurement environment.

The TSP is also a very useful signal in the measurement of very long impulse responses of rooms. The details of the measurement methods using the time-stretched pulse are discussed in references [1, 2].

3.7 Deformation of Impulse Response

Consider reexamining Figs. 3.3, 3.4 and 3.6 that compared the estimated impulse responses. In order to estimate impulse responses by the cross-spectrum method, the following three methods were used in those examples:

(1) N-point rectangular window applied to the input and output (i.e. no weighting),
(2) N-point Hanning window applied to the input and output, and
(3) N-point rectangular window and additional N-point zero padding applied to the input and a $2N$-point rectangular window applied to the output.

In all three figures, the impulse responses obtained by methods (1) and (2), especially by (2), seem to have faster decays than the true impulse responses. It is important to know why the estimated impulse responses are deformed and how this deformation is related to the window.

3.7.1 Deduction of the Deformation Function

One reason for the deformation of the impulse response is the leakage of the response from the specific window. Another reason is the change in the amplitude envelope of the response due to the weighting by the window.

It is assumed that the window length N is equal to or longer than the impulse response. Figure 3.11 shows the transfer system and its relationship to the windowed input and output. The input $x(n)$ is given to the input terminal and the output $y(n)$ appears at the output terminal. The windowed input $w(n)x(n)$ and windowed output $w(n)y(n)$ are used when the impulse response is obtained by the cross-spectrum method.

Figure 3.12 shows data from the impulse response (a), the input and output sequences (b and c), and the system response to the m-th input sample. It is assumed that the impulse response exists from 0 to $N - 1$ with constant amplitude. The input sequence after the windowing, $u(n) = w(n)x(n) \quad n = 0,\ 1,\ \ldots,\ N$, is shown by the open vertical bars and the m-th sample, $u(m) = w(m)x(m)$ is shown by the filled vertical bar in Fig. 3.12b. This is the sequence that is observed at terminal (3) in Fig. 3.11. Figure 3.12c shows the sequence that is observed at terminal (4).

Figure 3.12d shows how the response to the input sample $x(m)$ is shortened and deformed by the windowing. The response that appears at terminal (4) to the input sample $x(m)$ is given by Fig. 3.12. Deformation of the response to the m-th input sample by the time window. (a): impulse response, (b) windowed input sequence, (c): windowed output sequence, (d): response to the m-th input impulse.

$$v(n) = w(n)x(m)h(n - m).$$

Therefore, the response that appears at terminal (4) to the unit impulse at the discrete time m is given by

$$h_m(n) = w(m + n)h(n) \tag{3.20}$$

This is valid only for the impulse which is input at the discrete time m, and since the window length is N, it becomes zero when $m + n \geq N$ or $n \geq N - m$.

Then, the response $u(n)$ at terminal (3) and the response $v(n)$ at terminal (4) to the input $x(m)$ are given, respectively, by

$$u(n) = w(m)x(m)\delta(n - m) \tag{3.21}$$

$$v(n) = \begin{cases} x(m)h_m(n - m) & n \geq m \\ 0 & n < m \end{cases} \tag{3.22}$$

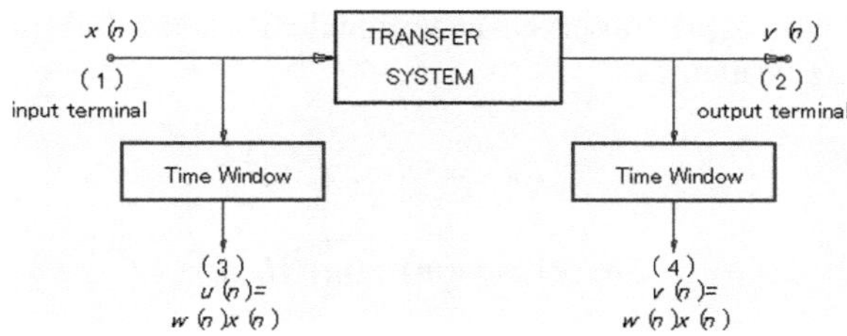

Fig. 3.11 Input and output of a transfer system and their signal terminals with time windows

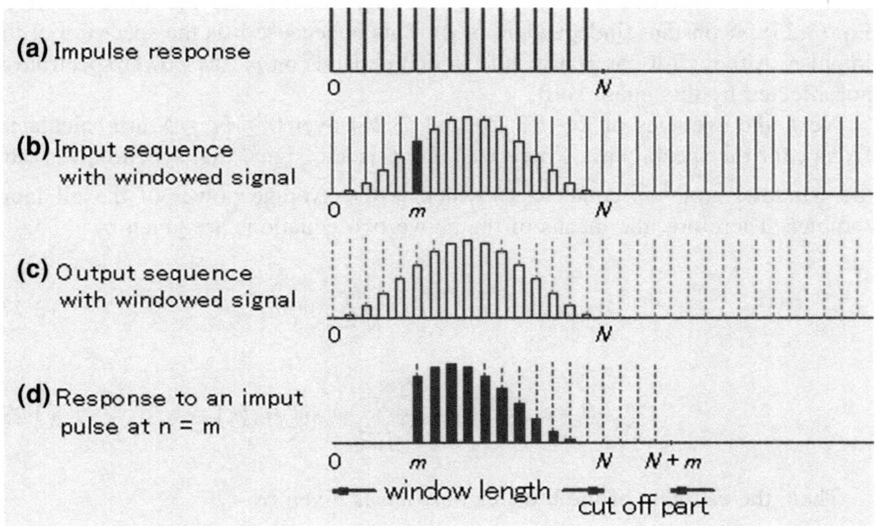

Fig. 3.12 Deformation of the response to the m-th input sample by the time window **a** Impulse response (constant value of length $N - 1$), **b** Windowed input sequence, **c** Windowed output sequence, **d** Response to the m-th input impulse

The spectra of these sequences obtained by N-point DFT are given by

$$U(k) = w(m)x(m)\exp\left(-j2\pi\frac{km}{N}\right) \tag{3.23}$$

$$V(k) = w(m)H_m(k)\exp\left(-j2\pi\frac{km}{N}\right) \tag{3.24}$$

where $H_m(k)$ is the N-point DFT of $h_m(n)$.

Using the above relationship, the power spectrum $S_{UU}(k)$ of the m-th input (not only for one $w(m)x(m)$ but for many other $w(m)x(m)s$ that are observed in individual N-point sequences for many times of averaging) and the cross-spectrum

$S_{UV}(k)$ of the $w(m)x(m)$ (observed at terminal (3)) and $w(m)y(m)$ (observed at terminal (4)) are written as

$$S_{UU}(k) = w(m)^2\overline{x(m)^2} \tag{3.25}$$

$$S_{UV}(k) = w(m)^2\overline{x(m)^2}H_m(k) \tag{3.26}$$

where $x(m)$ is the m-th sample of each input sequence and $\overline{x(m)^2}$ is the mean of $x(m)^2$ after many averages.

It may be superfluous to say that the magnitude of the spectrum given by Eq. (3.25) is constant (independent of k). This is because it is the spectrum of the impulse. Although it has phase shift, which depends on m, the power spectrum is not affected by this phase shift.

Next, the averages of Eqs. (3.25) and (3.26) over $0 \le m \le N$ are calculated. Except for the special case where $x(n)$ is a periodic signal that synchronizes with the window, $\overline{x(m)^2}$ is equal to $\overline{x^2}$ which is the average power of the all input samples. Therefore, the means of the above two equations are given by

$$W_{UU}(k) = \overline{S_{UU}(k)} = \frac{\overline{x^2}}{N}\sum_{m=0}^{N-1} w(m)^2 \tag{3.27}$$

$$W_{UV}(k) = \overline{S_{UV}(k)} = \frac{\overline{x^2}}{N}\sum_{m=0}^{N-1} w(m)^2 H_m(k) \tag{3.28}$$

Then, the estimate of the transfer function is given by

$$\hat{H}(k) = \frac{W_{UV}(k)}{W_{UU}(k)} = \frac{\sum_{m=0}^{N-1} w(m)H_m(k)}{\sum_{m=0}^{N-1} w(m)^2} \tag{3.29}$$

The IDFT of Eq. (3.29) is the estimate of the impulse response. Since only $H_m(k)$ of the numerator of Eq. (3.29) is dependent on k and $H_m(k)$ is the DFT of $h_m(n)$ given by Eq. (3.30), the estimate of the impulse response $\hat{h}(n)$ is given by

$$\hat{h}(n) = \frac{\sum_{m=0}^{N-1} w(m)w(m+n)}{\sum_{m=0}^{N-1} w(m)^2} h(n) = D(n)h(n) \tag{3.30}$$

where $w(m)$ is the time window function and $D(n)$ is the deformation function of the impulse response obtained by the cross-spectrum method. The latter is explicitly expressed as

$$D(n) = \frac{\sum_{m=0}^{N-1} w(m)w(m+n)}{\sum_{m=0}^{N-1} w(m)^2} \tag{3.31}$$

By inserting various window functions $w(n)$ into Eq. (3.31), their deformation functions can be obtained.

3.7.2 The Deformation Functions of Actual Time Windows

By applying various window functions to Eq. (3.31), deformations of impulse responses due to various window functions are obtained. Those are shown here:

(1) Rectangular window

Window function:

$$w_R(n) = 1 0 \leq \mathrm{n} < \mathrm{N} \tag{3.32}$$

Deformation function:

$$D(n) = \left(1 - \frac{n}{N}\right) \tag{3.33}$$

(2) Hanning window

Window function:

$$w_N(n) = 0.5\left(1 - \cos\frac{2\pi n}{N}\right) \tag{3.34}$$

Deformation function:

$$D(n) = \frac{G(n) + H(n)/2\pi}{A^2 + \frac{B^2}{2} + \frac{C^2}{2}} \tag{3.35}$$

where,

$$G(n) = \left(A^2 + \frac{B^2}{2}\cos\frac{2\pi n}{N} + \frac{C^2}{2}\cos\frac{4\pi n}{N}\right)\left(1 - \frac{n}{N}\right)$$

$$H(n) = \left(2AB - \frac{B^2}{2} - \frac{2BC}{3}\right)\sin\frac{2\pi n}{N} - C\left(A - \frac{4B}{3} + \frac{C}{4}\right)\sin\frac{4\pi n}{N}$$

A = 0.5
B = 0.5
C = 0

(3) Hamming window

Window function:

$$w_M(n) = \left(0.54 - 0.46\cos\frac{2\pi n}{N}\right) \tag{3.36}$$

Deformation function:
Same as Eq. (3.35) with A = 0.54, B = 0.46, C = 0

(4) Blackman-Harris window

Window function:

$$w_B(n) = \left(0.423 - 0.498\cos\frac{2\pi n}{N} + 0.0792\cos\frac{4\pi n}{N}\right) \tag{3.37}$$

Deformation function:
Same as Eq. (3.35) with A = 0.423, B = 0.498, C = 0.0792

(5) Half sine window

Window function:

$$w_S(n) = \sin\frac{\pi n}{N} \tag{3.38}$$

Deformation function:

$$D(n) = \cos\frac{\pi n}{N}\left(1 - \frac{n}{N}\right) + \frac{1}{\pi}\sin\frac{\pi n}{N} \tag{3.39}$$

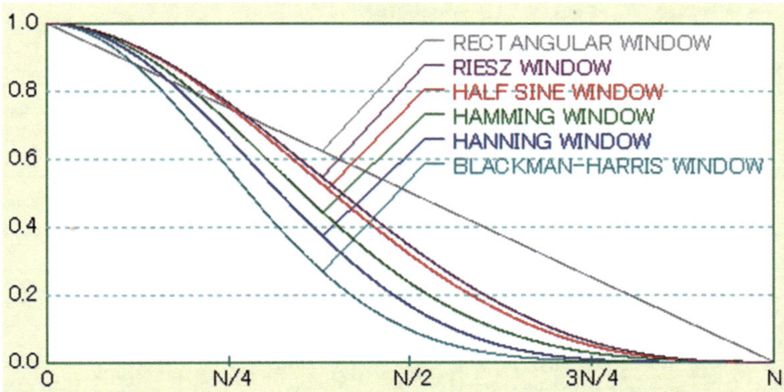

Fig. 3.13 Dependence of deformation function $D(n)$ on the window types when the impulse response is estimated by the cross-spectrum method

(6) Riesz window

Window function:

$$w_S(n) = 1 - \left(1 - \frac{2n}{N}\right)^2 \tag{3.40}$$

Deformation function:

$$D(n) = 5\left(2 - 3\frac{n}{N} + 2\frac{n^2}{N^2}\right)\left(1 - \frac{n}{N}\right)^3 - 9\left(1 - \frac{n}{N}\right)^5 \tag{3.41}$$

In order to visually show the degree of deformation, the numerical values of the deformation factors are shown in Fig. 3.13. This figure shows variations of the envelopes of the impulse responses. The envelope of the impulse response decreases linearly in the case of the rectangular window, but for the other windows, the degree of deformation is small in the range within initial 10 % of the window length. However, the amplitude of the impulse response reduces quickly as n/N increases beyond 10 %.

Since it is not clear how the computed impulse responses may change, the simulation results are shown in Fig. 3.14 for the case when the impulse response is an exponentially decaying sine wave which decays 20 dB linearly from $n = 0$ to N.

The window functions from the top are (1) Rectangular window, (2) Hanning window, (3) Hamming window, (4) Blackman-Harris window, (5) Half sine window and (6) Riesz window. The waveforms of the estimated impulse response are drawn in the left hand side and the squared amplitude is drawn in the right hand side in dB scale $\left(10\log(|h(n)|^2)\right)$.

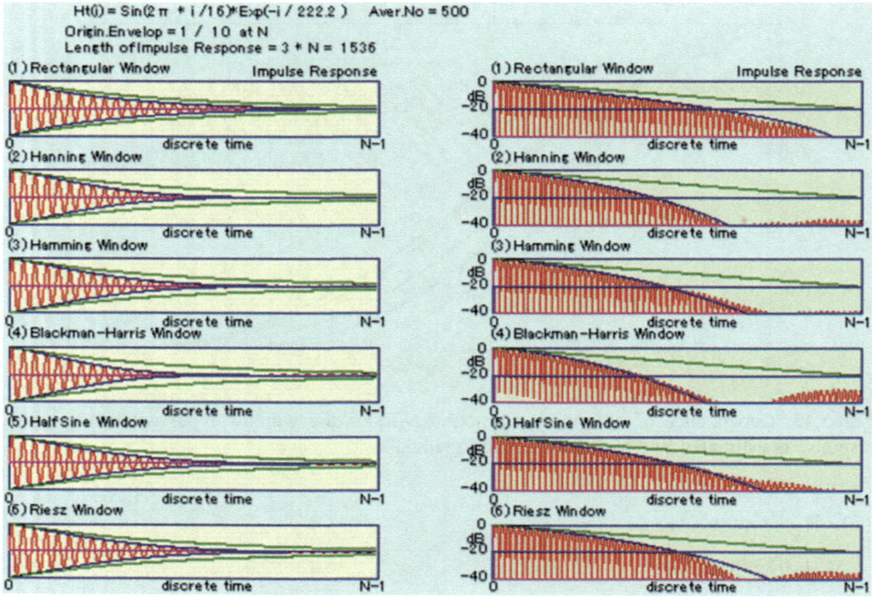

Fig. 3.14 The results of the computer simulation of the impulse response estimation by the cross-spectrum method. Animation available in supplementary files under filename E10-14_ZA4_Simulation.exe

The dotted lines show the true reduction rate of the impulse response and the dashed lines show the reduction rates including the theoretical deformation factors. The impulse responses are obtained by 500 averages. The deformations of the impulse responses agree very well with those obtained by the theory. Some of them show differences in the range with large reductions. These are made smaller if the number of averages is further increased.

At first glance, the rectangular window seems the best, due to the attenuation in the small amplitude range. But in most practical cases, the decrease of amplitude in the initial 1/8 of the sequence is important. The deformation in the initial 1/8 range is less for the Riesz and half sine windows; much more deformation is observed in the case of rectangular window. Those relations were demonstrated by Fig. 3.13.

Therefore, if the impulse response spans the full range of the window, i.e., if the decay is slow, the deformation becomes non-negligible. However, since the usual impulse response decays as a function of time, the reduction of amplitude in the later time of the response has a small effect on the residual power ratio. If the impulse response reduces exponentially, the deformation is negligible if the window length is longer than 8 time constants (the time constant is the time required for the amplitude of the impulse response to reduce to $1/e(\approx 0.368)$).

In the measurement of decay processes in general and reverberation times in architectural acoustics, in which the determination of decay rate is crucial, the influence of the window function and its length should not be ignored. In order to avoid these problems, the method described in the next Sect. 3.7.3—an extension of method (3) in Sect. 3.4—is recommended. The method using the TSP signal introduced in Sect. 3.6 is also useful.

3.7.3 Impulse Response Estimation Without Amplitude Deformation

Methods that do not produce amplitude deformations in the impulse response estimation have already been demonstrated in Figs. 3.9 and 3.10. Using the method described in Sect. 3.6, a response without amplitude deformation can be achieved within the range up to $n = N$, if a periodic signal with period N is used as an input signal. The method (3) introduced in Figs. 3.3, 3.4 and 3.6 is also useful: a pair of N-point input sequences (N-point input with N-point zeros added) and $2N$-point output sequence are sampled and the $2N$-point DFT/IDFT are applied to them. Furthermore, as described in Sect. 3.6, the cause of amplitude deformation principally does not exist if the TSP signals are used, without depending on the cross-spectrum method.

In Sects. 3.7.1 and 3.7.2, it was shown that amplitude deformation is produced if N-point input and output sequences and N-point DFTs are used. It was also shown that amplitude deformation is negligible if the length of the window (=N) is much longer than the length of the impulse response. However, since this information alone may be misunderstood, a related method will be explained next in order to show how to determine the length of the minimum window length that avoids amplitude deformation.

It is obvious that the condition $N > L$ is necessary in order to estimate the impulse response with length L using the N-point cross-spectrum method.

The larger the value of N *is*, the smaller will be the amplitude deformation. On the other hand, N should be as small as possible for computational efficiency. After choosing an appropriate value for N, the length M of the input sequence $x(n)$ is varied with the constraint $M = N - L$, as shown in Fig. 3.15. The remaining length L in the input is replaced by zeros. The full N-point output sequence $y(n)$ is used and the impulse response is estimated using the N-point cross-spectrum method. This method assures that the first $(N - M)$ points of the impulse response data are not deformed.

However, since the response corresponding to the input during $M < n < N$ effectively works as noise in $y(n)$, a large number of averages is necessary to reduce the effect of the noise, particularly if M is small and L is relatively large compared to N.

Figure 3.16 shows the results of impulse response estimation by this method. An infinite random sequence is input to the system, and the system impulse response is a constant amplitude sine wave with length N (=512). Since the

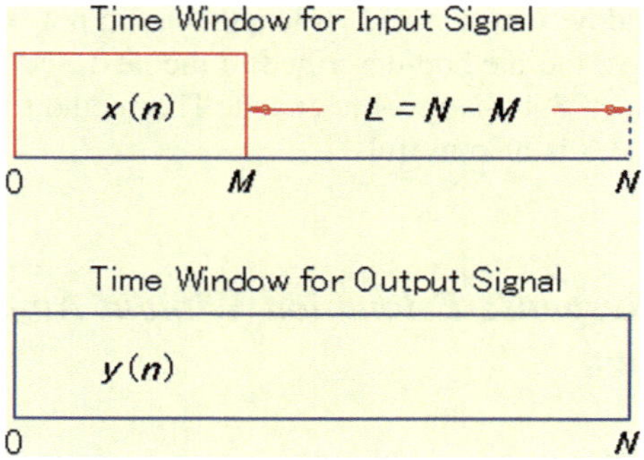

Fig. 3.15 Determination of window length in order to avoid amplitude deformation

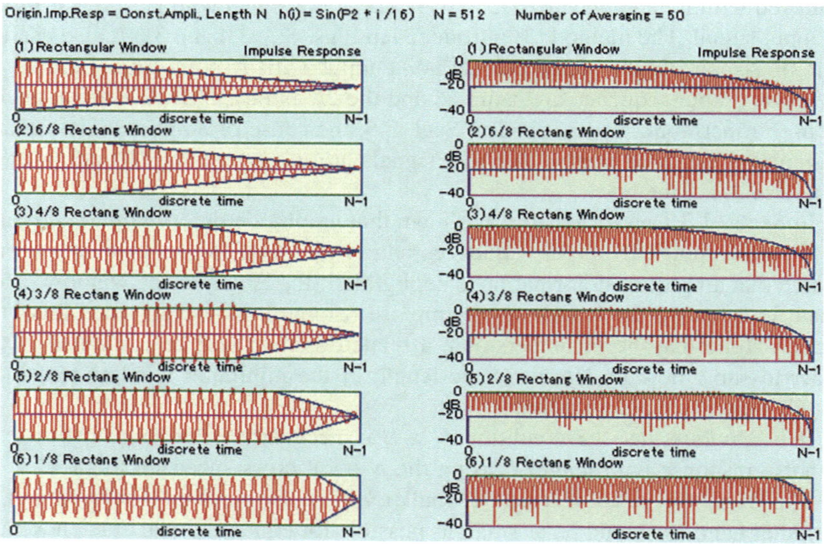

Fig. 3.16 Impulse responses estimated by the method shown in Fig. 3.15. The input is a random sequence, and the output is calculated as a convolution of the input and the impulse response. The latter is a constant amplitude sine wave with length *N*. The vertical axes of the *left-side* and *right-side* charts are in linear and dB scales, respectively. Animation available in supplementary files under filename E10-16_ZA4_Simulation.exe

amplitude of the impulse response is constant, the reduction of the output amplitude identifies the amplitude reduction rate itself.

The charts from the top in Fig. 3.16 are for $M/N = 1$, 6/8, 4/8, 3/8, 2/8, 1/8, respectively. The vertical axes are shown in linear and dB scales in the left-side and right-side charts, respectively. The broken lines in the charts trace the estimated envelopes of the impulse responses. Note that they are constant (no deformation) in the range $0 \leq n \leq N - M$, and decay with constant slope in the left-side (linear scale) charts and with curvature in the right-side (dB scale) charts.

3.8 Coherence Function

In the previous sections, it was shown how to estimate the transfer functions and impulse responses by the cross-spectrum method using the input and output of transfer systems.

It was shown that noise in the output that is not dependent on the input is eliminated by averaging many times (if the noise is large compared with the output, the number of averages must be increased). The purpose of this section is to develop an index of the degree of noise contamination in the output signal.

Examine Fig. 3.1, which shows the transfer system and the relationship between the input and the output. The signal $x(t)$ is input to a transfer system with impulse response $h(t)$. The observed output $y(t)$ is a summation of the convolution, $v(t)$, of the input $x(t)$ and the impulse response $h(t)$ and the external noise $n(t)$. The expressions in the time and frequency domains are given by Eqs. (3.5) and (3.6), respectively.

Consider Eq. (3.42) below, which is Eq. (3.6) repeated.

$$Y(f) = V(f) + N(f) = X(f)H(f) + N(f) \tag{3.42}$$

If there is no external noise, the output spectrum is $V(f)$, which has the following relation with $X(f)$

$$V(f) = X(f)H(f)$$

The above equation shows that $V(f)$ is proportional to the input $X(f)$, with the transfer function $H(f)$ defining the proportionality and delay at each frequency. Both the amplitude and phase relationships between the input and the output are determined by the transfer function. If $x(t)$ is a swept sine wave, interference patterns appear due to the phase changes of the input and output when they are superimposed. In this sense, it is said that there is a *coherency* between $x(t)$ and $v(t)$.

On the other hand, the noise $n(t)$ is completely independent of $x(t)$, and the amplitude and phase relationships always change between every sample record. No interference patterns are observed even when they are superimposed.

Since $v(t)$ has a coherency with $x(t)$ and $n(t)$ has no coherency with $x(t)$, the ratio $[v(t)/y(t)]$ may be a choice to express the degree of coherence. Unfortunately, this ratio of time-domain functions cannot be a meaningful quantity. Since the interference is an indication of the phase changes at the same frequencies, the degree of interference should be expressed as a function of frequency. For this purpose, the ratio of the power spectra is another choice. However, only the signals at terminal (1) and (2) can be observed and not the output signal $v(t)$ (i.e. $V(f)$). It is necessary to find a power spectrum ratio of $v(t)$ to $y(t)$ from the observable signals $x(t)$ and $y(t)$.

The power spectrum of $v(t)$ is given by

$$V^*(f)V(f) = X^*(f)H^*(f)X(f)H(f) = X^*(f)X(f)H^*(f)H(f)$$

Using the definitions,

$$W_{XX}(f) = X^*(f)X(f)$$

$$W_{VV}(f) = V^*(f)V(f)$$

the above equation can be rewritten as

$$W_{VV}(f) = W_{XX}(f)H^*(f)H(f) \tag{3.43}$$

On the other hand, it is known from Eq. (3.11) that $H(f)$ is given by the ratio of the cross-spectrum of the input and output to the input power spectrum. Substituting Eq. (3.11) into Eq. (3.43), the following relationship is obtained.

$$W_{VV}(f) = W_{XX}(f)\frac{W_{XY}^*(f)}{W_{XX}(f)}\frac{W_{XY}(f)}{W_{XX}(f)}$$

The ratio of the power-spectrum $W_{vv}(f)$ of $v(t)$ to the power-spectrum $W_{YY}(f)$ of $y(t)$ has the dimension of the square of the amplitude ratio. We will denote that ratio by γ_{XY}^2.

$$\gamma_{XY}^2(f) = \frac{W_{VV}(f)}{W_{YY}(f)} = \frac{|W_{XY}(f)|^2}{W_{XX}(f)W_{YY}(f)} \tag{3.44}$$

Equation (3.44) is calculated using only input and output signals. It is the ratio of the power of the output component caused only by the input to the power of the total output. For this reason, γ_{XY}^2 is called the *coherence function* between $x(t)$ and $y(t)$.

Equation (3.42) through (3.44) were derived using input and output spectra given as continuous functions of frequency. What are actually needed are equations that are based on discrete sample sequences of input and output signals. For this purpose, the expressions using continuous frequencies must be replaced with ones using discrete frequencies. As has been mentioned so many times, the spectra used in Eq. (3.44) are Fourier transforms of time waveforms that range from $-\infty$ to $+\infty$, which are different from the periodograms obtained by the DFT.

Commence the process from Eqs. (3.45) and (3.46) in the discrete time domain.

$$y(n) = V(n) + n(n) = x(n) * h(n) + n(n) \tag{3.45}$$

$$Y(k) = V(k) + N(k) = X(k)H(k) + N(k) \tag{3.46}$$

An operation similar to that used in deriving Eq. (3.44) will be performed. First of all, attention to the relationship between these two equations should be observed.

In general, the Fourier transform of the convolution is given by the product of the individual spectral functions. If the Fourier transform is replaced by the DFT, circularity becomes an issue in the discussion.

In order for

$$V(k) = X(k)H(k) \tag{3.47}$$

to be an exact DFT of

$$V(n) = X(n) * h(n) \tag{3.48}$$

all variables must be periodic sequences of period N. This cannot always be satisfied in reality. Keeping this in mind, the periodogram of $v(n)$ is derived from Eq. (3.47) as

$$P_{VV}(k) = V^*(k)V(k) = X^*(k)X(k)H^*(k)H(k) \tag{3.49}$$

It is known that the average of the N-point power spectrum calculated by the DFT does not necessarily produce a correct power spectrum. But, it is also known from the discussion in Sect. 2.7 that errors will not be significant unless conditions are exceptionally poor. The ratio of the averages of the N-point periodograms of $v(n)$ and $y(n)$ is given by Eq. (3.49):

$$\overline{P_{VV}(k)} = \overline{V^*(k)V(k)} = \overline{X^*(k)X(k)}H^*(k)H(k) \tag{3.50}$$

Substituting $H(k)$ given by Eqs. (3.18), (3.51) is obtained.

$$\overline{P_{VV}(k)} = \overline{V^*(k)V(k)} = \frac{\overline{X^*(k)Y(k)}^*\,\overline{X^*(k)Y(k)}}{\overline{X^*(k)X(k)}} \tag{3.51}$$

Denoting the ratio of $\overline{P_{VV}(k)}$ and the periodogram of $y(n)$ by $\gamma_{XY}^2(k)$, Eq. (3.51) is obtained.

$$\gamma_{XY}^2(k) = \frac{\left|\overline{X^*(k)Y(k)}\right|^2}{\left|\overline{X(k)}\right|^2\left|\overline{Y(k)}\right|^2} = \frac{\left|\overline{P_{XY}(k)}\right|^2}{\overline{P_{XX}(k)}\cdot\overline{P_{YY}(k)}} \tag{3.52}$$

This is the equation for the coherence function in the discrete domain.

The maximum value of the coherence ($\gamma_{XY}^2(k) = 1$) indicates that the output is completely caused by the input. As the coherence approaches unity, the impulse response estimation becomes more reliable. Since coherence is a function of frequency, it also provides information about which frequencies may have measurement problems.

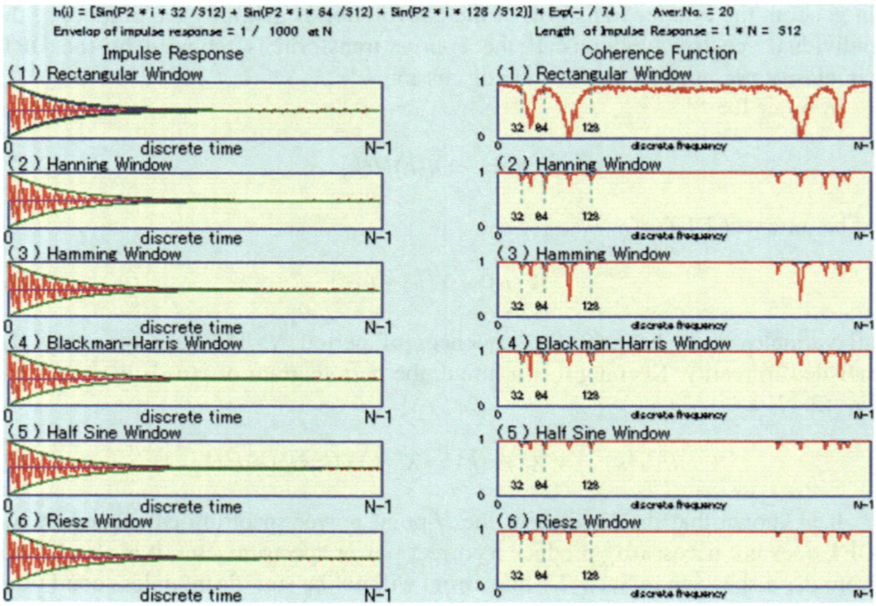

Fig. 3.17 Dependence of impulse response and coherence functions estimated from the input and output of a transfer system on the types of windows. Animation available in supplementary files under filename E10-17_COHERENCE_Simulation.exe

Properties of the coherence function using another transfer system model will now be examined. The impulse response of the system is the summation of three decaying sine waves, and a random sequence is used as the input signal. Results of the coherence calculated with six different types of windows are shown in Fig. 3.17.

The coherence with the rectangular window exhibits wide and deep troughs. The dips of the coherence function for other windows, however, appear only at or near the impulse response frequencies and they are much shallower.

In order to understand the reasons for the large coherence reduction by the rectangular window, the dependence of the coherence on the ratio of the impulse response length to the window length was investigated for the rectangular and Hanning windows. The results are shown in Fig. 3.18. The length of the impulse response is defined here as the time at which the amplitude decreases to 1/1,000 from the maximum amplitude at $t = 0$ (defined as the reverberation time in architectural acoustics).

Figure 3.18 shows that, if the impulse response length is short compared with the window length, the reduction of coherence is small. In the case of the rectangular window, if the impulse response length exceeds 1/4 of the window length, a non-negligible reduction is observed. On the other hand, in the case of the Hanning window, the reduction of coherence remains small even as the length

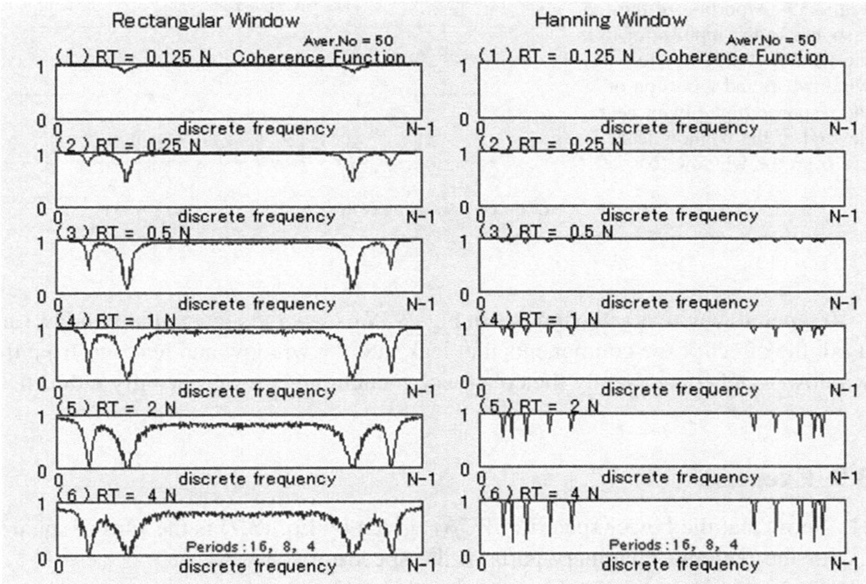

Fig. 3.18 The effect of the impulse response length (reverberation time) on the estimated coherence functions. From the *top*, the lengths are doubled from *N*/8 to 4*N*. The *left-side* and *right-side* charts are for the rectangular and Hanning windows, respectively. Animation available in supplementary files under filename E10-18_CoHAN.exe

ratio approaches 1. Even when the ratio exceeds 1, the reduction is limited to several narrow frequency regions.

The reason the coherence decreases is that the long decay of the resonance leaks out from the window. The sharp endings on both sides of the waveform cut out by the rectangular window widen the distribution of frequency components and consequently the bands of coherence reduction. The results with other tapered windows are similar to those of the Hanning window. This can be checked by running the program of Fig. 3.17.

The coherence reduction for the rectangular window is exceptionally large compared to the other windows. The major distinction between the rectangular window and the other windows is that the rectangular window is the only window not tapered at both ends. Because of this, the response to the input just before the front side of the window leaks directly into the window (see Fig. 3.19a). Since this component has nothing to do with the input signal in the window, it reduces the coherence. If the impulse response is long, the power that leaks into the window is relatively large and the reduction of the coherence is also large. Similarly, the response to the input just before the closing of the window leaks out of the window (see Fig. 3.19b) and works to reduce the coherence.

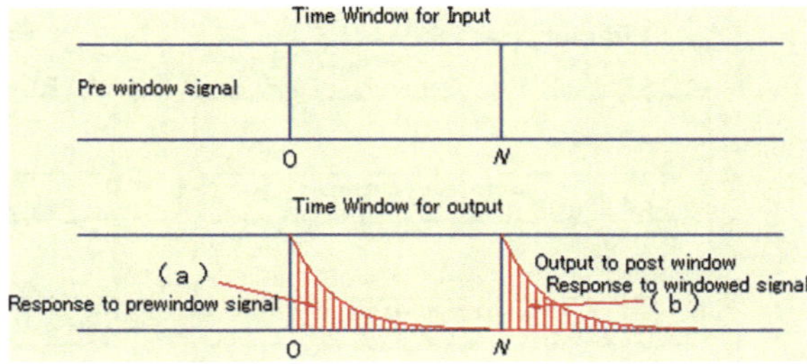

Fig. 3.19 A portion of the response to the input prior to the window falls into the window (**a**) and a portion of the response to the input near the end of the window leaks out from the window (**b**)

If tapered windows (all windows in Fig. 3.17 except the rectangular window) are used, the effect of the components that leak into the window and leak out from the window—and consequently the coherence reduction—are significantly reduced.

3.9 Exercise

1. Show that the power spectrum $W_{XX}(f)$ given by Eq. (3.7) is the sum of squares of the real and imaginary parts of the spectrum.
2. Show that Eq. (3.7) is always real and Eq. (3.8) is complex in general.
3. What kind of relationship exists between the power spectrum and the auto-correlation function?
4. What kind of relationship exists between the cross-spectrum and the cross-correlation function?
5. What is the condition for the cross-spectrum of $x(t)$ and $n(t)$ to be zero?
6. What is the transfer function?
7. What is the IDFT of the product of the N-point DFT of the input and the N-point transfer function?
8. What is the requirement for the product of the DFT of the input and the transfer function to be the DFT of the output without noise?
9. What is the average of the N-point cross-periodogram of the input and output divided by the average of the N-point periodogram of the input?
10. What is the common property of the impulse response obtained from the cross-spectrum of the input and output when a non-random (correlated) sequence is used as an input?
11. Is averaging of the DFT necessary when the impulse response is estimated from the cross-spectrum of a random input and an output under a condition without noise?
12. What is the necessary condition(s) for the precise estimation of the impulse response from the cross-spectrum of a random input and an output?
13. Is averaging of the FFT necessary when an impulse response is estimated by an N-point cross-spectrum method using an N-point periodic input signal?
14. A signal is mixed with several time-independent frequency components and a random noise, where the latter is uncorrelated to the former. Describe a method of obtaining frequency components by eliminating the noise.

Chapter 4
Cepstrum Analysis

We will discuss here a somewhat strange method, which is called *cepstrum analysis*. The cepstrum is defined as a Fourier transform of the logarithm of spectrum. There is a wider general mathematical theory that covers the theory of the cepstrum referred to as homomorphism. However, we will take a conventional approach without stepping into the realm of mathematics. Our main interest thus far has been on investigating information of waveforms by applying the Fourier transform. There is no reason to limit applying the Fourier transform to waveforms. For example, if a spectrum has periodicity, a line spectrum or line spectra will be found by the Fourier transform. It will be a clue for clarifying the generation mechanism of a waveform, reason(s) why the periodicity exists, and eventually, the observed phenomenon itself.

4.1 Propagation Time Difference and Spectrum

Let us consider the case of an observed signal with direct and reflected waves as shown in Fig. 4.1. What we need to know is the propagation time difference between the two waves. This analysis technique was started from the study in the petroleum exploration engineering. The depth of the reflecting layer under the ground can be known from the propagation time of the elastic wave emitted from the surface into the ground. Figure 4.1 shows the frame format for this case. It is assumed that there are two waves observed at the observation point. They can be considered either the direct and reflected waves or waves with two different propagation paths.

The direct wave is denoted by $x(n)$ and the same signal is reflected by a discontinuous surface and arrives at the observation point with the time delay d and the r times of amplitude modification. The time delay d and the reflection coefficient r are generally frequency dependent but it is assumed here that it is frequency independent for simplicity. At the observation point, the direct and the reflected waves are not separated and observed as one waveform, which is denoted

K. Kido, *Digital Fourier Analysis: Advanced Techniques*,
DOI: 10.1007/978-1-4939-1127-1_4,

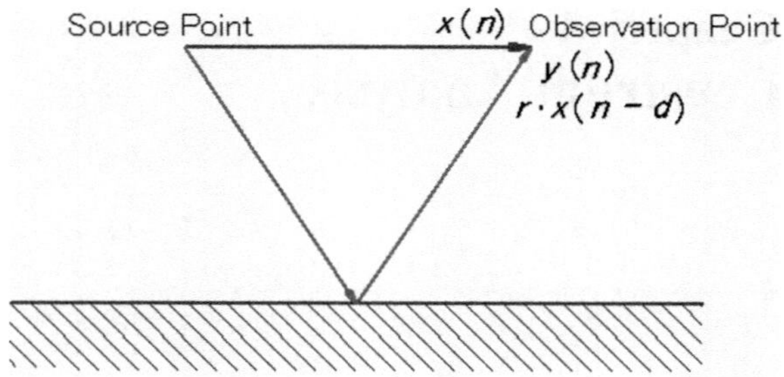

Fig. 4.1 Combined waveform $y(n)$ with a direct wave $x(n)$ and a reflected wave $r{\cdot}x(n-d)$ with reflection coefficient r and time delay d between them

by $y(n)$ in Fig. 4.1. The task here is to determine d from the observed signal $y(n)$. Under this condition, $y(n)$ is expressed by

$$y(n) = x(n) + rx(n-d) \tag{4.1}$$

If $x(n)$ is a single impulse, $y(n)$ is composed of two impulses separated by d, which can be easily determined. If $x(n)$ is a continuous signal, the situation is not so simple. Figure 4.2a, b show $x(n)$ and $y(n)$, respectively, when $x(n)$ is a random sequence generated by a computer. The reflection coefficient is 0.7 and the delay d has the time length of 32 samples. We need to determine d from Fig. 4.2b, but it is even impossible to tell that (a) and (b) are related to each other.

From the discussion in Chap. 2 of vol. 1, we know that the spectrum of a delayed signal has a phase delay equal to the product of the time delay and the frequency. The spectra of $x(n)$ and $x(n-d)$ have the same amplitude and the latter has extra phase shift than the former. Therefore, the spectrum of $y(n)$, the summation of the two, should have something periodic on the frequency axis.

If $X(k)$ is a DFT of $x(n)$, then the DFT of $x(n-d)$ has the phase delay which is proportional to the product of time delay d and the frequency k as shown below (see Appendix 4B)

$$\mathrm{DFT}[x(n-d)] = X(k)\exp\left\{-j2\pi\frac{dk}{N}\right\}$$

From this relation, the DFT of Eq. (4.1) becomes

$$Y(k) = \mathrm{DFT}[y(n)] = X(k)\left\{1 + r\exp\left(-j2\pi\frac{dk}{N}\right)\right\} \tag{4.2}$$

The right-hand side of Eq. (4.2) has a complex exponential function which varies as a function of k with period N/d. Therefore, $Y(k)$ should have something periodical. Once the period is known, the time delay d can be also known.

In order to see the dependence of the magnitude on the frequency, let us take the power spectrum (periodogram) of Eq. (4.2), which is the square of the absolute value of $X(k)$.

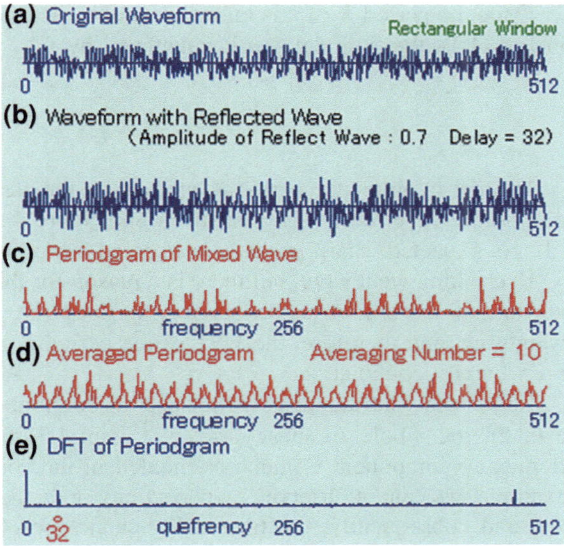

Fig. 4.2 Example of periodogram and its DFT of a random sequence added with the same signal but with 0.7 times of amplitude change and 32-point sample delay. Animation available in supplementary files under filename E11-02_Ceps.exe

$$\begin{aligned} P_{YY}(k) &= Y^*(k)Y(k) = |X(k)|^2\left\{\left(1 + r\cos 2\pi\frac{dk}{N}\right)^2 + r^2\sin^2 2\pi\frac{dk}{N}\right\} \\ &= P_{XX}(k)\left\{1 + r^2 + 2r\cos 2\pi\frac{dk}{N}\right\} \end{aligned} \tag{4.3}$$

In the periodogram above, the last term in {} has the amplitude change with a constant period as a function of k. Figure 4.2c shows this periodicity.

$P_{XX}(k)$ of the right-hand side of Eq. (4.3) is a periodogram of a finite sequence extracted from the signal source, which is dependent on the property of the source. If $x(n)$ is a random sequence, it is expected that $P_{XX}(k)$ stays constant without regard to k. On the other hand, the term inside {} is a function of only the reflection coefficient r and the time delay d, and independent of the property of the signal source.

As mentioned before, the signal source is a random sequence generated by a computer. This assures that $P_{XX}(k)$ is almost constant ("almost" since the length of the analyzed sequence is not infinite). It has been also mentioned that the reflection coefficient r is 0.7 and the delay time d is equal to the 32-point sample time.

Figure 4.2a is the original sequence, (b) is the addition of two waves (Eq. 4.1), and (c) is the periodogram without averaging. Figure 4.2d is the periodogram after 10 times of averaging. Since the signal source is random, it is expected that the effect of averaging may not be significant, but the periodicity of the spectrum becomes clearer.

Once the period K of k is known, since this is the period of the cosine function in Eq. (4.3), the time delay d is calculated by

$$d = \frac{N}{K} \tag{4.4}$$

In order to obtain the period K, we can use the idea that the Fourier transform of a periodic waveform has a line spectrum. Figure 4.2e is the Fourier transform of (d). As expected, sharp peaks are observed at points 32 and 480 (=512−32).

Everything went right, but there is a reason for this. Since the signal source is a portion of a random sequence, $P_{XX}(k)$ does not have any specific frequency weighting, and therefore, the DFT of the periodic function in { } of Eq. (4.3) has been luckily obtained.

Let us see what happens if the spectral distribution of the signal source is not even in the whole frequency range. Figure 4.3 shows a case where the high-frequency component is much attenuated in the source signal. As shown in this example, we cannot observe any periodicity in the periodogram of the mixed wave (b), and consequently, the time delay cannot be obtained from the DFT of the periodogram.

Furthermore, as we have learned in Chap. 2, the DFT of Eq. (4.3) is actually not more than the auto-correlation function of $y(n)$ (except for the difference of magnitude) calculated indirectly from the power spectrum.

4.2 DFT of Logarithmic Periodogram

The right-hand side of Eq. (4.3) is the product of $P_{XX}(k)$ and the periodic component. In the case of Fig. 4.2, the signal source is a random sequence and $P_{XX}(k)$ does not have any specific weighting on the k axis, and therefore, the time delay is easily obtained. What was actually done was to obtain the auto-correlation from the power spectrum. Cases in which $P_{XX}(k)$ is largely dependent on k, like the case of Fig. 4.3, must be taken into account.

In the case of Fig. 4.3, the frequency components are very small in the high-frequency range and almost zero except for the very low frequency range. It is hard to observe the periodicity in the power spectrum. Equation (4.3) indicates that if the periodicity exists in the frequency range where $P_{XX}(k)$ is very small, its effect on the total spectrum is very small. This is why the line spectrum does not appear in the DFT of the power spectrum. In Chap. 6 of vol. 1 (Appendix 6A), it was shown that the decibel scale (logarithm) can be used in order to make small values visible.

There is another important property of the logarithm. The logarithm of multiplication of two terms is given the summation of logarithms of individual terms. The logarithm of Eq. (4.3) is given by the summation of logarithm of $P_{XX}(k)$ and logarithm of { }. Even if $P_{XX}(k)$ gets smaller as k increases, its logarithm decreases

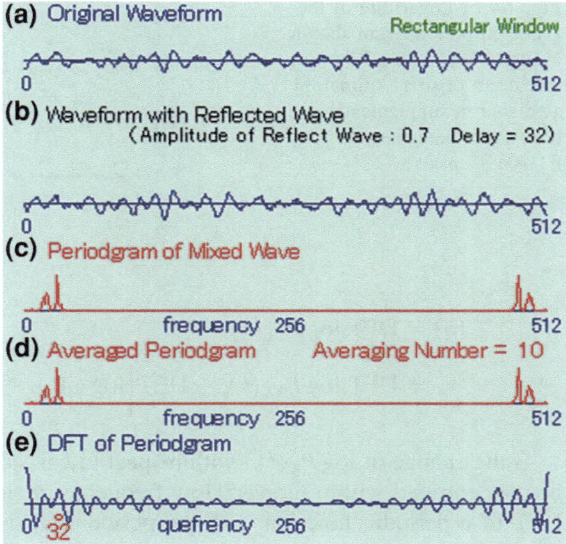

Fig. 4.3 Example of periodogram and its DFT of a low-frequency weighted sequence added with the same signal but with 0.7 times of amplitude change and 32-point sample delay. Animation available in supplementary files under filename E11-03_Ceps.exe

simply in the negative direction as k increases. The logarithm of { } varies periodically as k increases. By taking the logarithm, the periodicity of { } is not lost by the multiplication of small values of $P_{XX}(k)$.

Let us check this by equations.

$$\log P_{YY}(k) = \log P_{XX}(k) + \log\left\{1 + r^2 + 2r\cos 2\pi\frac{dk}{N}\right\} \tag{4.5}$$

The first term of the right-hand side of this equation is the logarithm of the periodogram, which decreases in the negative direction as k increases in the example of Fig. 4.3. But this change is separated from the second term. The second term is a wavy function which varies periodically. These two properties are not lost even they are added. If the DFT is applied to the logarithm of the power spectrum, the periodicity will show up.

Let us check if this is the case. The logarithm of the power spectrum of Fig. 4.3d is shown in Fig. 4.4a. The embossment in the low-frequency region corresponds to the spectrum shown in Fig. 4.3c. In this figure, the vertical axis is in logarithmic scale and the invisible small values in the high-frequency region in Fig. 4.3c and d are now clearly seen. The DFT of the logarithmic periodogram clearly shows the line spectra at $n = 32$ and 480 even though the periodicity is not as clear as the one in Fig. 4.2.

Now, let us apply the DFT to $\log P_{YY}(k)$ of Eq. (4.5) assuming that k is a discrete time. It becomes as follows.

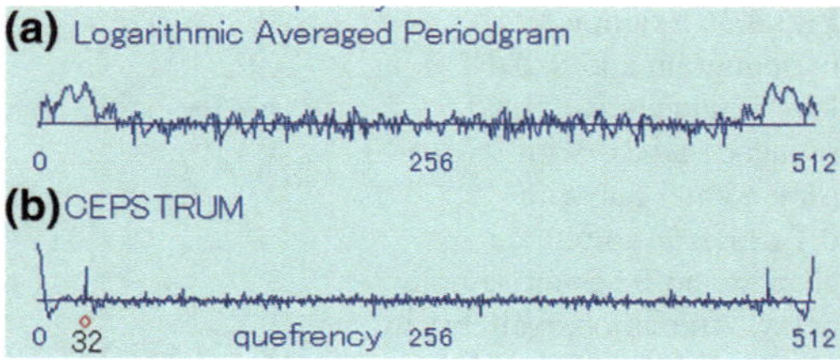

Fig. 4.4 **a** Logarithm of the averaged periodogram shown in Fig. 4.3d, and **b** DFT of (e) (cepstrum of (d)). Animation available in supplementary files under filename E11-04_Ceps.exe

$$\begin{aligned} C(q) &= \mathrm{DFT}[\log P_{YY}(k)] \\ &= \mathrm{DFT}[\log P_{XX}(k)] + \mathrm{DFT}\left[\log\left\{1 + r^2 + 2r\cos 2\pi\frac{dk}{N}\right\}\right] \end{aligned} \quad (4.6)$$

If the change of $\log P_{XX}(k)$ with respect to k is slow, its DFT, the first term, will be concentrated within the very low frequency region. Since the second term is a DFT of a periodic function with the period N/d, there will be a peak at its reciprocal, d/N.

In Fig. 4.4, the DFT of (e) is given by (f). There is a large peak at the left end and another sharp peak at 32 and 480 (512−32) which correspond to the first and the second terms of the right-hand side of Eq. (4.6), respectively.

If the horizontal axis of Fig. 4.4a is time, then the axis of its DFT should be frequency. The horizontal axis of Fig. 4.4a is frequency, and its IDFT has time as its horizontal axis. Since Fig.4.4b is the DFT of (e) and we cannot say that its horizontal axis represents time. Because of this reason, the inventors of this theory, Bogert et al. [3] introduced for this horizontal axis a new word *quefrency* by exchanging "fre" and "que" of frequency. Since Fig.4.4b is the DFT of a spectrum, they also introduced a new word *cepstrum* for it by inverting the order of "spec." Another word they invented is *lifter* as the weighting function on the quefrency axis instead of the filter on the frequency axis. Among many other new words, cepstrum and quefrency are commonly used. The processing using the cepstrum is referred to as *cepstrum analysis.*

Since the quefrency corresponds to the frequency of a spectrum but its dimension is time, small and large quefrencies correspond to short and long times, respectively. Therefore, the adjectives short and long are used for the quefrency.

This method was invented by trying to find the times needed for the elastic wave that was generated on the ground surface and reflected back by the layers under the ground. Figure 4.5 summarizes the process of cepstrum analysis shown separately in Figs. 4.2, 4.3, and 4.4. However, the source signal used in Fig. 4.5 is a short sequence extracted using a 100-point Hanning window. In this method, since it is possible to take care of more than one reflections, two reflected waves with 32- and 94-point sample delays are analyzed.

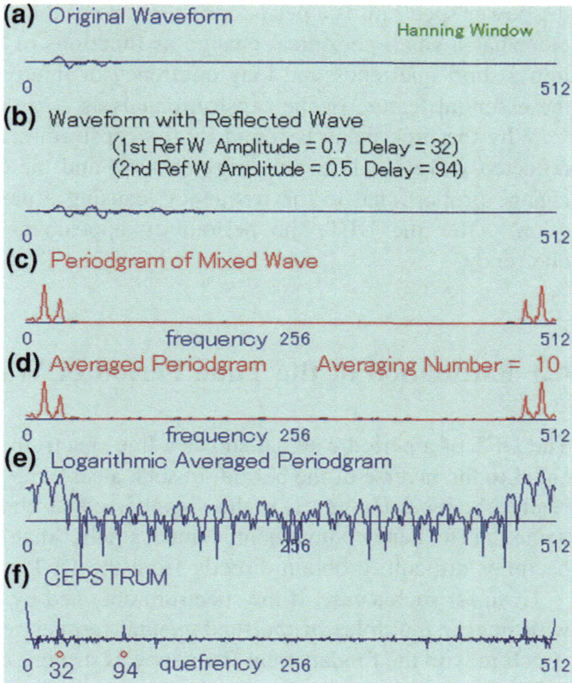

Fig. 4.5 Cepstrum analysis process of estimating delay times from a signal composed of low-pass filtered 100-point long random sequence and its 32- and 94-point delayed signals. Animation available in supplementary files under filename E11-05_Ceps.exe

The signal used in the example of Fig. 4.5 is short and its spectrum distribution is in a limited area. However, two reflected waves are clearly seen in at 32 (480) and 94 (418) in Fig. 4.5f indicating that the cepstrum analysis is very effective.

Let us review the process of the cepstrum analysis with Fig. 4.5.

(a) original waveform
(b) combination of the original and reflected waveforms.
(first reflection: amplitude 0.7, delay 32)
(second reflection: amplitude 0.5, delay 94)
(c) N-point periodogram of (b) ($N = 512$ in the present case)
(d) Averaging of (c) (may improve to enhance the periodicity but not always necessary)
(e) Calculation of the logarithmic power spectrum
(f) DFT of the logarithmic periodogram (=cepstrum given as a function of quefrency)

As shown above, if a spectrum of a time function is represented as a product of two functions, it is converted to a summation of two terms by taking the logarithm and then the Fourier transform is applied to them. Since the waveform and power spectrum are represented by discrete sequences, the Fourier transform is DFT and the power spectrum is a periodogram. If one of the spectra changes slowly and the other has a small periodical change as functions of frequency, they are separated into a short quefrency and long quefrency or line quefrency components. This is the essential feature of the cepstrum analysis.

Why the time delay is found by the cepstrum analysis is that spectrum of the reflected wave is add to the original wave and the combined signal gets a phase change proportional to the frequency, causing a periodic change in the periodogram. After the DFT, the periodicity appears as a peak at a corresponding quefrency.

4.3 Estimation of the Time Period of Waveform

The DFT of a periodic signal shows a line spectrum at the fundamental frequency equal to the inverse of the period. In such a case, the period of the signal should be easily obtained. However, if the signal is contaminated with a noise or the fundamental frequency component is much smaller than the higher order harmonics, it becomes difficult to obtain directly from the DFT.

Even for such a case, if the spectrum obtained by the DFT contains line spectra with integer multiples of the fundamental frequency, the spectrum has a periodic structure with the fundamental frequency as its period. In order to obtain its period, the frequency axis of the spectrum is considered to be the time axis so that the DFT can be applied. The result, cepstrum introduced in the previous section, will give the periodicity of the signal.

The sine (or cosine) wave is an extreme of periodic waves. In this case, however, the cepstrum analysis cannot be applied, since the sine wave has only one spectrum line (in the positive frequency range) and it does not have the periodicity. The auto-correlation will give the periodicity, but the auto-correlation of a sine wave is also a sine wave. In these cases, it is better to obtain the fundamental frequency directly from the power spectrum. Signals best fit for the cepstrum analysis are those which contain many harmonics and also their spectra have periodicity along the frequency axis. The cepstrum analysis may be applied, for example, to machine noises and sounds made by animals with many high-frequency components. Sometimes, external noises may be mixed with them.

Here, the following signal is used for the demonstration. A computer-generated random sequence is 3-point moving averaged. Then a 130-point sequence is extracted from the original sequence and a periodic signal is produced by sequentially connecting the identical 130-point sequences. Also a random noise sequence with a 50 % amplitude is added. Figure 4.6 shows the results of the cepstrum analysis. The 1,024-point Hanning-windowed signal is shown in Fig. 4.6a. We cannot see any periodicity from this signal.

There are two peaks in the periodogram in Fig. 4.6b obtained by 1,024-point DFT. The lower peak is approximately at $1024/130 = 8$. However, there is a larger peak with a slightly higher frequency. Also, even higher frequency components

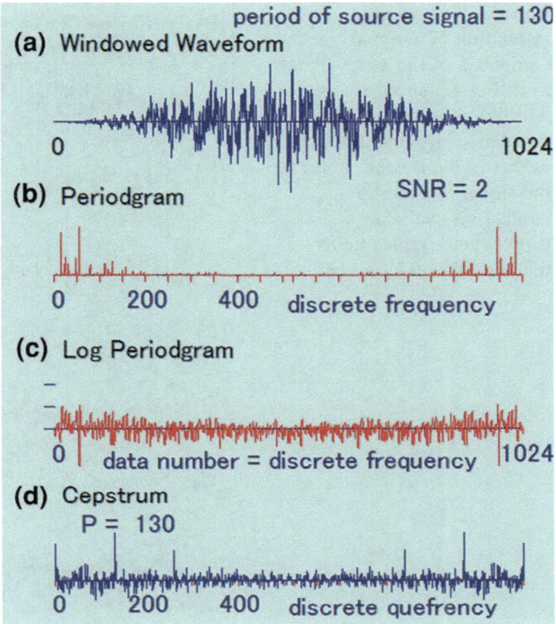

Fig. 4.6 Periodicity estimation by the cepstrum analysis. A signal used here is a periodic sequence made of 130-point long 3-point moving-averaged random sequence added with a random noise. The latter has an amplitude of 1/2 of the former. Animation available in supplementary files under filename E11-06_Cep.exe

exist with smaller amplitudes. From these facts, it is difficult to consider that the original signal has a periodicity. In this case, the periodic signal contains a noise of approximately 1/2 of amplitude, but it is difficult to tell that this is a periodic signal even without the noise. You will see it by running the program in the CD.

Figure 4.6c, the logarithmic periodogram, makes small components visible, but still it is difficult to say that this signal is periodic. Figure 4.6d, however, clearly shows a peak at quefrency 130. This is exactly the period used when the signal was made.

What happens if a sine wave is added to this signal? It is a common experience in a real measurement that a measured signal contains a component of the electric power frequency.

Keeping such a case in mind, we will add a sine wave with period 80 and 50 % amplitude of the original periodic sequence to the signal of the previous example. Results obtained by the same process as in Fig. 4.6 is shown in Fig. 4.7. The waveform seems to have a periodicity with roughly 80, but the peak you get from the cepstrum is only at quefrency 130. The example of this case shows that an addition of a sine wave does not produce any effect on the result of the cepstrum analysis.

In the above example, since the amplitude of the added sine wave is much larger than other frequency components, its spectrum appears as the largest peak in the periodogram. After taking the logarithm (by logarithmically compressing the

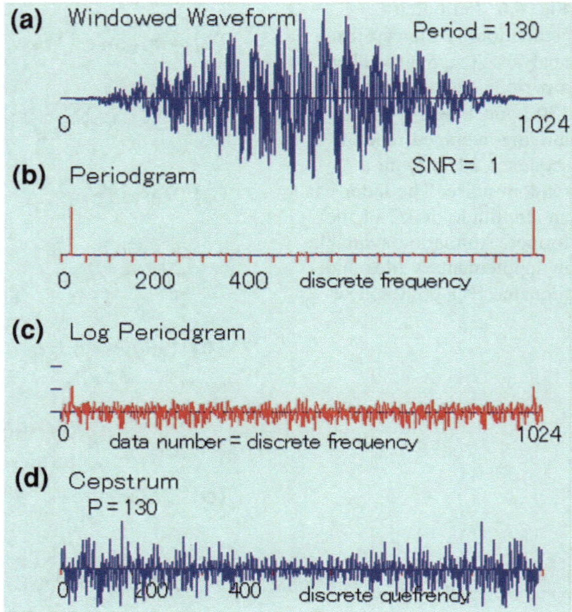

Fig. 4.7 Periodicity estimation of a signal composed of the same signal as in Fig.4.6 and a sine wave with 80-point period. The latter has the same amplitude as that of the random noise in the signal of Fig. 4.6. Animation available in supplementary files under filename E11-07_Cep.exe

spectrum), it becomes compatible with other components. As a result, the added sine wave becomes negligible in the cepstrum analysis. This is the reason why Fig. 4.7d is almost identical with Fig. 4.6d.

The programs of Figs. 4.6, 4.7, and 4.8 (in the next section) let you perform the DFT without taking the logarithm of the power spectrum. You can compare the results of the cepstrum analysis with those of the auto-correlation analysis. You will see that the cepstrum analysis is more advantageous when a signal is contaminated with a periodic noise as shown in Fig. 4.7.

4.4 Application of Period Estimation

Let us take a speech waveform as another example that has periodicity in the spectrum. The vocal sound generated by the vibration of the vocal chord, which is excited by breathing, is characterized by the acoustical filter called vocal tract from the vocal chord to the mouth and lips. The percentage of the vowels in the spoken Japanese words is very high and they are categorized into five groups, /a/, /i/, /u/, /e/, and /o/. The waveform of the vocal sound is similar to a triangle wave and contains multiples of harmonics. The transfer function of the vocal tract has three or four resonances in the frequency band from 100 to 3,000 Hz, and changes smoothly as a

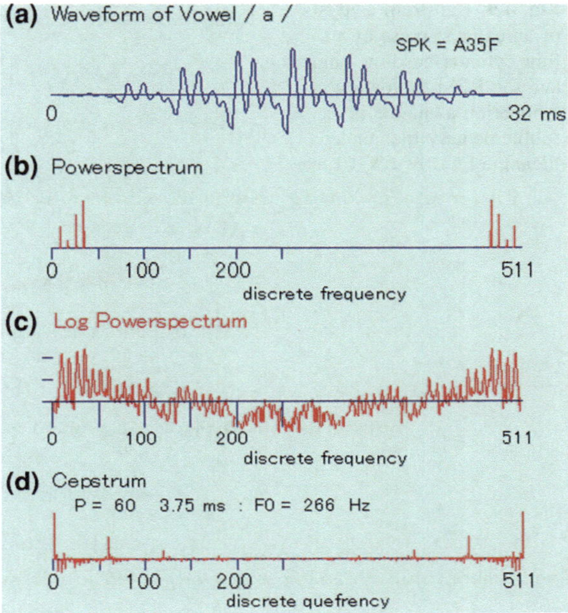

Fig. 4.8 Estimation of pitch period of a male vowel /a/ by the cepstrum analysis. Animation available in supplementary files under filename E11-08_Vowel.exe

function of frequency. The resonance frequencies are referred to as the 1st, 2nd, and 3rd formant frequencies in the ascending order.

Let us apply the cepstrum analysis to the vowels. Results are shown in Fig. 4.8. The 16-kHz sampling frequency and the 512-point DFT (i.e., 32 ms window length) are used for the analysis. In this example, the estimated pitch period is 60, which corresponds to 3.75 ms (=60/16). The fundamental frequency of 267 Hz is given by the reciprocal of the pitch period.

The peak in Fig. 4.8 that gives the pitch period is very small. The reason for this is the fluctuation of the pitch. You can check the effect of the fluctuation by adding fluctuation in the order of ± 2 using the program of Fig. 4.7.

There is a large peak at a very short quefrency in Fig. 4.8d. This is the same with Fig. 4.7. Since the pitch of speech is almost always within 2–10 ms, the peak in the quefrency range shorter than 2 ms is not a pitch period. It has another meaning, but it will be discussed later.

Before that, a more complex example will be shown. It is an estimation of rpm (revolution per minute) from exhaust noises of engines. Figure 4.9 shows the results but some explanations are necessary since the calculation of the cepstrum is slightly different from the previous examples. It also needs some knowledge about engines for the estimation of the rpm.

Figure 4.9a is the waveform of an exhaust noise extracted using the 8,192-point Hanning window with the 48-kHz sampling frequency. The window length is 170.67 ms (=8192/48000). Figure 4.9b shows the lowest 256-point data of the

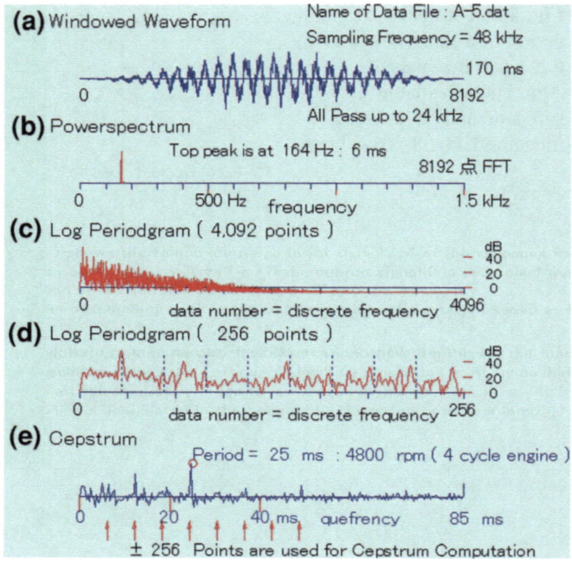

Fig. 4.9 Cepstrum analysis of an exhaust noise of a four-cylinder gasoline engine and the RPM estimation. Animation available in supplementary files under filename E11-09_EX_01.exe

power spectrum obtained by the 8,192-point FFT. Figure 4.9b shows the largest peak at 158 Hz, which may be directly estimated from the waveform of (a). Other components are too small to be seen in this figure. The discrete frequency corresponding to 158 Hz is 27 (=158 × 0.17067). The second largest peak is at 404 Hz (discrete frequency at 69). We can guess that the combustion occurs 158 times per second (9,480 times per minute) but this is not enough to estimate the rpm (revolution of the axle per minute). We need to know how many cylinders the engine has and also the engine is of the two-cycle or four-cycle type. Since, in this case, the engine is a four-cylinder and four-cycle type, the rpm is calculated by 9480 × 2/4 (=4740).

In the four-cycle engine, each cylinder has one combustion in each two cycles. Therefore, if the period of the combustion is obtained, the rpm will be known.

As shown in Fig. 4.9c, the level of the high-frequency components is low and easily influenced by the external noise, and therefore, the high-frequency components are not suitable for the period estimation. Figure 4.9d shows the details of the power spectrum in the discrete frequency range from 0 to 256 (1.5 kHz). The 512-point DFT of the power spectrum from −255 to 256 gives the cepstrum shown by Fig. 4.9e. It is possible of course to apply the whole 8,192-point DFT to the logarithmic power spectrum of (c), but it is sometimes more beneficial to remove the high-frequency components, if it is contaminated with an unnecessary noise.

In the lower part of Fig.4.9e, the upward arrows are shown that indicate the quefrency 6.3 ms, which is the reciprocal of 158 Hz, and its integer multiples. They are named as the 1st, 2nd, …, respectively, in the ascending order. The 4th and 8th peaks are dominant compared to other peaks, among which the 4th peak at 25 ms is the largest. This indicates that 25 ms is the period of the whole cycle of the engine. The peak at 1/4 of 25 ms corresponds to the period between neighboring combustions. The whole cycle of the engine is the cycle in which every cylinder finishes one combustion. In this example, since the period of 25 ms, which is four times of 6.3 ms, is the whole cycle, it is known that the engine has four cylinders. This means that rps is 80 (=2/0.025). This is why 4,800 rpm is written in Fig.4.9e. You can see various results by changing the rpm and conditions to calculate the cepstrum.

4.5 Estimation of a Gross Pattern of a Transfer Function

As mentioned before, the vowels are the results of characterization of the roughly triangular waveform of the vocal sound by the transfer functions of the vocal tract. By denoting the vocal sound by $x(n)$ and the impulse response of the vocal tract by $h(n)$, the vowel $v(n)$ is obtained by the convolution of the vocal sound and the impulse response, which is given below.

$$v(n) = x(n) * h(n) \tag{4.7}$$

By applying the DFT on both sides, we have

$$V(k) = X(k)H(k) \tag{4.8}$$

where the convolution in the time domain is expressed by the product in the frequency domain.

Figure 4.8 shows how the pitch is obtained when $v(n)$ is observed. For the cepstrum analysis, the logarithm of the power spectrum is taken.

$$\log[V(k) * V(k)] = \log[X(k) * X(k)] + \log[H(k) * H(k)] \tag{4.9}$$

By using the absolute symbol | |, it can be rewritten as

$$\log|V(k)| = \log|X(k)| + \log|H(k) \tag{4.10}$$

The above equation shows that the convolution of two functions in the time domain is expressed by the summation of logarithms of their absolutes in the frequency domain.

In Fig. 4.8, Eq. (4.10) is given by (c). It shows that the spectrum of the vocal sound is composed of the fundamental and harmonic components and has fine up and downs like a comb. This period is obtained as the quefrency of the cepstrum peaks shown by Fig. 4.8d. Then, what corresponds to the second term $\log|H(k)|$ of Eq. (4.10) must be the general gradual change of the logarithmic spectrum given by Fig. 4.8c.

By taking the DFT of Eq. (4.10) following the previous example, the comb-like fine undulation and the gradual change can be separated as follows:

$$\begin{aligned} C(q) &= \mathrm{DFT}[\log|V(k)|] \\ &= \mathrm{DFT}[\log|X(k)|] + \mathrm{DFT}[\log|H(k)|] \end{aligned} \tag{4.11}$$

This is exactly the cepstrum. The first term of the right-hand side of Eq. (4.11) is determined only by the signal source (vocal sound in the speech) and the second term is determined only by the transfer function. If these two are separated in the different regions of the quefrency axis, it becomes possible to separate them.

Look at Fig. 4.8d, which is the DFT of (c). The peak at quefrency 113 is due to the fine periodic undulation of the power spectrum, which corresponds to the first term of Eq. (4.11). The other peaks concentrated in the region near quefrency 0 are due to the transfer function of the vocal tract. Since the change with respect to quefrency is gradual, the main cepstrum components are concentrated near quefrency 0.

As we have seen above, the first and second terms in Eq. (4.11) are separated in the quefrency axis. The former corresponds to the sharp peaks in the long quefrency range and the latter corresponds to the components in the shorter quefrency range, respectively. If we let the peaks and longer components be zero, the remaining cepstrum only has the components related to the transfer function of the vocal tract.

$$C_H(q) = \mathrm{DFT}[\log|H(k)|] \tag{4.12}$$

If this is the case, the IDFT of Eq. (4.12) will give $\log|H(k)|$, the amplitude characteristic of the transfer function.

The above idea was applied to the example of Fig. 4.8. The results are shown in Fig. 4.10. Since the voice sample and the process until the cepstrum is obtained is the same as in Figs. 4.8 and 4.10 starts from the cepstrum. The cepstrum has a peak at quefrency 113 (real time 7.06 ms since the sampling frequency is 16 kHz), which is a result of the comb-like undulation of the logarithmic power spectrum.

The components in the range shorter than this peak quefrency will be the one given by Eq. (4.12). The IDFT of these components should be equal to log|H(k)|. However, it was found that the setting of the cutoff quefrency just below the peak quefrency causes many small ripples in the power spectrum. Therefore, the cutoff quefrency was set at 1/2 of the peak quefrency, and components in the longer quefrency range were removed (1/2 of the peak quefrency was experimentally determined). You can check how results change when the cutoff quefrency is varied by running the program of Fig. 4.10. The operation of the rectangular weighting on the cepstrum is referred to as "applying the rectangular quefrency window." Figure 4.10b shows the rectangular quefrency window by the dotted line and the windowed cepstrum by the solid line. The one-half of the cepstrum on the right-hand side is for the negative quefrency.

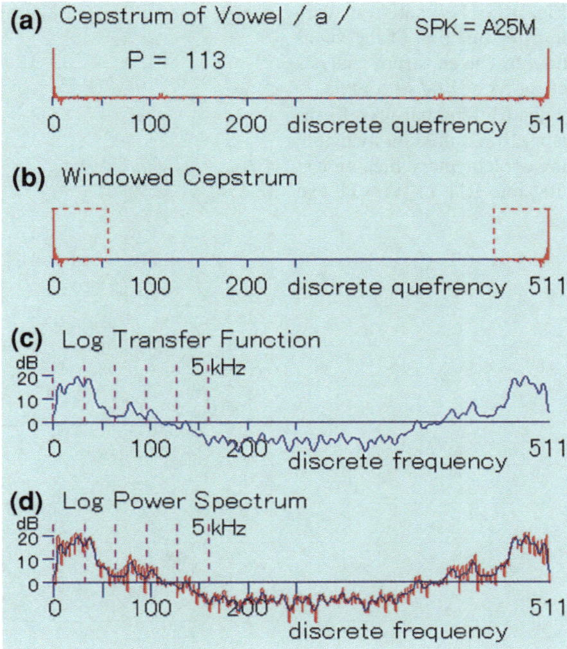

Fig. 4.10 Estimation of the transfer function of a vocal tract by use of a male voice /a/. Animation available in supplementary files under filename E11-10_VowTF.exe

The results of the IDFT are shown in Fig. 4.10c. This is the gross pattern of the transfer function estimated by the present method. Compared to the logarithmic power spectrum shown in Fig. 4.10d (same as Fig. 3.8c), Fig. 4.10c is a much smoother curve than Fig. 4.10d. If Fig. 4.10b represents Eq. (4.12) exactly, Fig. 4.10c, the IDFT of Fig.4.10b, should be the transfer function of the vocal tract (in the decibel scale).

However, speech scientists will say Fig. 4.10c is wrong. If it is a transfer function of the vowel /a/, it should have the first two peaks around 1,000 Hz and the peak of the third formant at around 2,000 Hz. Since the real frequency is given by multiplying the discrete frequency by 31.25, Fig. 4.10c has peaks approximately at 200, 800, 1050, 2500, and 3200 Hz. The peak around 200 Hz is too low for the formant frequency of /a/.

The reason for the generation of the peak in the low-frequency range is due to the large amplitude of the fundamental frequency component. By neglecting the lowest peak, it may be possible to relate the four peaks to the four formant frequencies.

It is not desirable to have a peak around 200 Hz in order to obtain the transfer functions of vocal tracts. Since the spectrum of the vocal sound has a general characteristic which is inversely proportional to the frequency, it is a common

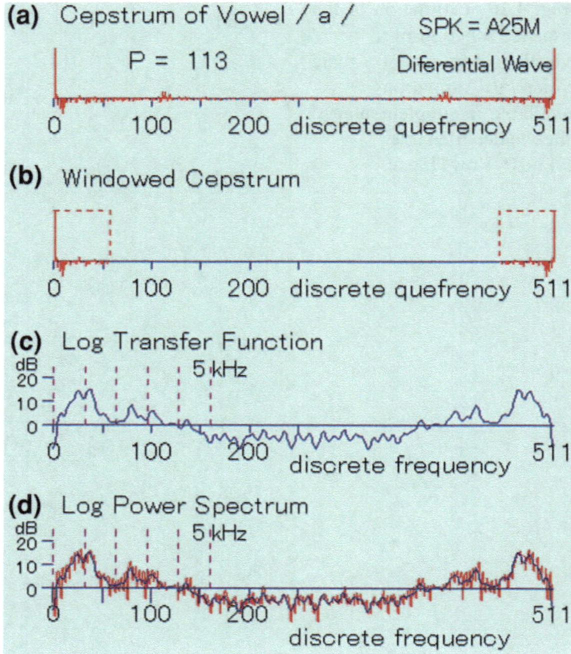

Fig. 4.11 Estimation of the transfer function of the vocal tract by the cepstrum analysis of the vowel /a/ of a male voice (differential waveform is used). Animation available in supplementary files under filename E11-11_VowTF.exe

practice in the speech analysis to use difference (differential) values of speech waveform to compensate for the level reduction in the high-frequency region. By taking the differential values, a chance of misunderstanding the fundamental frequency as the first formant is reduced.

Results obtained by the same process but using the differential waveform are shown in Fig. 4.4. It seems that a significant improvement of Fig. 4.11c from Fig. 4.10c has been achieved. However, the level in the low-frequency range still exists even though it has been lowered. Also there are too many ripples in the high-frequency range. One possible reason for this is the use of the rectangular quefrency window. It is clear from the discussion of Chap. 7 of vol. I that the use of a rectangular time window is not preferable in some cases in the spectrum estimation and a smoother window such as the Hanning window is more advantageous. For the same reason, the long quefrency components should be gradually attenuated by using a window such as the Hanning quefrency window.

Results obtained using the Hanning quefrency window is shown in Fig. 4.12. Since the Hanning window has the effective length of 1/2 of the rectangular window, the length of the Hanning window is twice of the rectangular windows in Figs. 4.10 and 4.11, i.e., the Hanning window becomes 0 at the peak quefrency. The results clearly show that the peak at 200 Hz is perfectly removed, and it seems that we can determine the formant frequencies choosing the peaks from the lowest one.

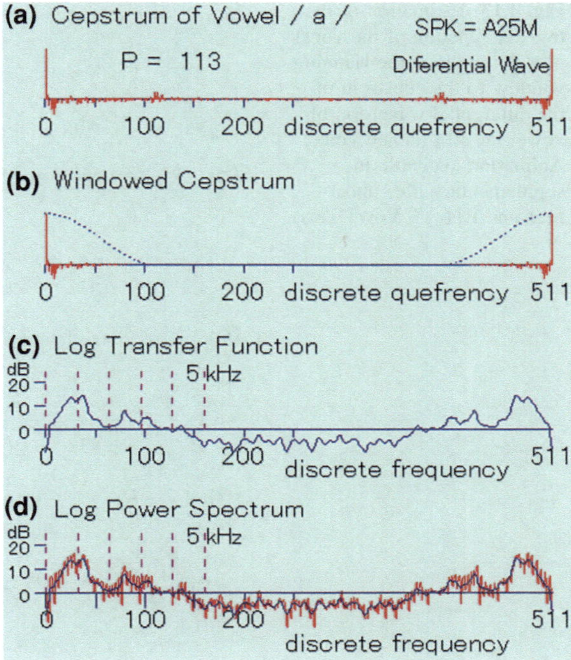

Fig. 4.12 Estimation of the transfer function of the vocal tract by applying the Hanning window to the cepstrum of the differential waveform of a vowel /a/ of a male voice. Animation available in supplementary files under filename E11-12_VowTF.exe

In the case of male voices, the fundamental frequency is below 150 Hz, which is smaller than the spacing between the adjacent formant frequencies, and therefore, it is mostly possible to obtain the formant frequencies by the cepstrum analysis. However, since the fundamental frequencies of female voices are mostly above 200 Hz, and sometimes close to 400 Hz, it is not appropriate to assume that the spectrum of the speech approximates the transfer function of the vocal tract.

In most of cases, it is difficult to obtain the formant frequencies of female voices by the cepstrum analysis. For example, as shown in Fig. 4.13, a dip is not observed between the 1st and the 2nd formants and they cannot be separated. This can be slightly improved by doubling the window length to 1,024-point long (64 ms).

Figure 4.13 is one of very successful cases and it is difficult to say that the cepstrum analysis is a good method to obtain the formant frequencies. The situation is worse in the real conversation. Therefore, in the analysis of the speech, modified versions of the cepstrum analysis are used [4]. Identification methods using vocal tract models with some resonance frequencies are also used [5]. Since they are outside of the Fourier analysis, they will not be treated here. Interested readers should refer to specialized books [6]. Since the cepstrum analysis is a general method which is independent of types of generic models, there are many areas of applications in the real world.

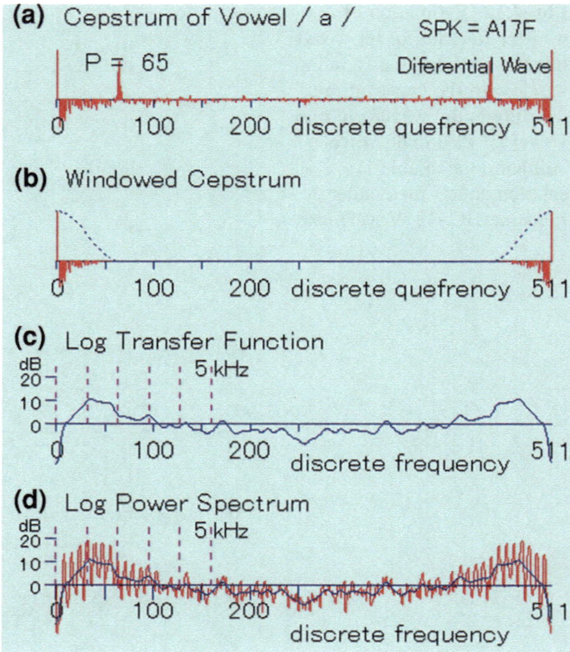

Fig. 4.13 Estimation of the transfer function of the vocal tract by applying the Hanning window to the cepstrum of the differential waveform of a vowel /a/ of a female voice. Animation available in supplementary files under filename E11-13_VowTF.exe

Thus far, we used spoken speech waveforms and therefore we cannot judge how accurate the transfer function estimations are. If we use a synthesized waveform, this problem can be avoided. Figure 4.14 shows the results.

Figure 4.14a shows the impulse response of the transfer system, (b) shows the absolute value of the transfer function (DFT of (a)), and (c) shows the logarithmic transfer function. Figure 4.14d is the input periodic waveform to the system, and (e) is the differential waveform of the output. Figure 4.14f and below show the results of the cepstrum analysis. Figure 4.14f is the logarithmic power spectrum, and (g) is the DFT of (f), which is the cepstrum. The horizontal axes of these two figures are doubled (only the positive frequency range is shown.). By applying the Hanning window to the cepstrum shown in (g), and then by applying IDFT to the result, the estimated logarithmic transfer function is obtained as shown in Fig.4.14h. Figure 4.14i is the result for the case without differentiation of the waveform.

Ideally, the estimated transfer function should be identical to (c). Figure 4.14h seems to be a good estimation of (c), but not identical. The result is good because the input signal is a triangular wave and the differentiation of the output is used before the cepstrum calculation. As this example shows, if the input is triangular and the differentiation of the output is used, the output becomes close to that of the impulsive input. If only one impulse exists, the spectrum is flat and a correct transfer function is obtained. But since the input is an impulse “sequence,” the

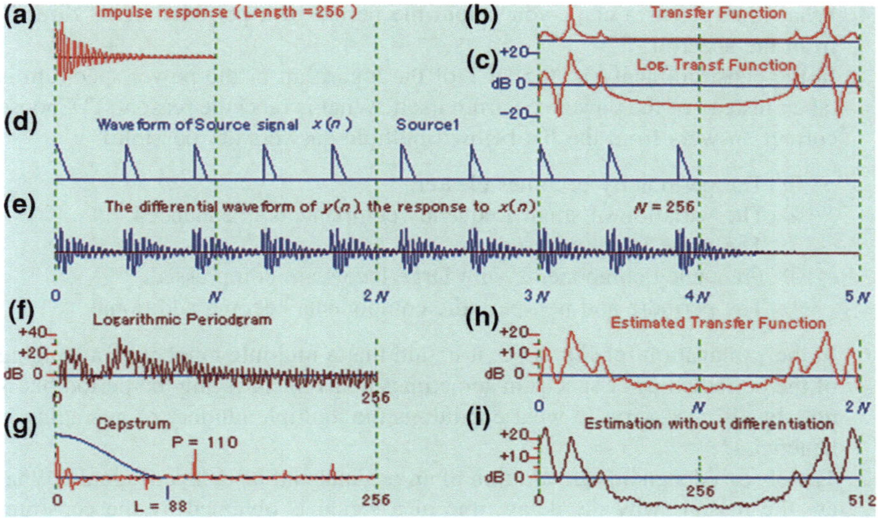

Fig. 4.14 Simulation of transfer function estimation by the cepstrum analysis. Animation available in supplementary files under filename E11-14_CEP-TF.exe

estimated result has a small irregularity. If the differentiation is not used, the level decreases roughly proportionally to the frequency. This is the effect of the input waveform. On the other hand, if the input is an impulse train, the differentiation should not be taken.

As we have seen, the estimation of the gross pattern of the transfer function by the cepstrum method is possible but not excellent. In the cepstrum analysis of the output of a transfer system driven by a periodic input, the applicable quefrency range gets narrow and fine structures of the transfer function cannot be recovered, which is principally unavoidable.

In many cases it is difficult to estimate transfer functions with sufficient accuracy by the cepstrum method. But since there are cases to which only the cepstrum analysis is applicable, it is worthwhile to fully understand this method and correctly use it.

4.6 Exercise

1. What kind of method to estimate the delay time d is possible if you also can measure the signal near the source in Fig. 4.1?
2. The DFT of a signal x(n) is X(k). If the signal is delayed by d, what is the DFT of the delayed signal?
3. If the additional sine wave has a period of 32 in Fig. 4.7, what is the discrete frequency of the line spectrum that will appear in Fig. 4.7b?

4. What kind of cases can't you obtain the period of a periodic signal directly from the spectrum?
5. In the cepstrum analysis, the DFT of the logarithm of the power spectrum is taken instead of the power spectrum itself. What is (are) the reason(s)? Choose correct answers from the list below (multiple answers are possible).

   ① The periodicity becomes clearer.
   ② The variation of small frequency components is enhanced.
   ③ The noise is compressed.
   ④ Frequency components with large levels are compressed.
   ⑤ The periodic and non-periodic components become additional.

6. In the explanation for Fig. 4.5d, it is said that a multiple number of averaging of the periodic power spectrum sometimes helps in clarifying the period but is not always necessary. In what conditions the multiple number of averaging is beneficial?
7. The observed signal is an addition of an original and its delayed signals. What are the reasons why the delay time of a signal is obtained by the cepstrum analysis?
8. Describe the method of obtaining the period of a signal by the cepstrum method.
9. Describe the process of obtaining the general shape of a transfer function by the cepstrum method.
10. What kind of input signal to the transfer system is preferable for the cepstrum analysis in estimating the general form of the transfer function?

Chapter 5
Hilbert Transform

First, we will discuss a problem of extracting the envelope of a sine waveform when its amplitude changes slowly as a function of time. If the waveform can be represented by a rotating vector and the vector length changes slowly as the time goes, the time change of the vector length is the envelope of the waveform. In order to represent the waveform as a rotating vector, we need another waveform that has a $\pi/2$ (90°) phase difference with the original waveform. Using the orthogonal relationship between the sine and cosine waveform, we will consider a general method of determining the orthogonal waveform to an original waveform. By further pursuing this method, we will end up with the Hilbert transform.

We will see that the real and imaginary parts of a transfer function that satisfies the causality satisfy the relationship of the Hilbert transform pair.

By expressing the waveform by a rotating vector, the instantaneous phase and frequency that were discussed in Chap. 1 of vol. I can be obtained. This problem will be discussed here once again from the standpoint of the Hilbert transform, and then the extraction of pitch, the amplitude and frequency modulations will be discussed.

5.1 Envelopes of Cosine Waves

The Euler's formula below shows that cosine wave is the real part of the complex exponential function $\exp(j2\pi ft)$.

$$\exp(j2\pi ft) = \cos(2\pi ft) + j\sin(2\pi ft) \tag{5.1}$$

This is the same with Eq. (1.5) in Chap. 1 of vol. I. The real part is the cosine wave and the imaginary part is the sine wave that has the $\pi/2$ (90°) phase delay. That is, the latter is orthogonal to the former or vice versa.

This function Eq. (5.1) is a rotating vector on the complex plane as shown in Fig. 1.10 in Chap. 1 of vol. I. Even a more general waveform that changes its

K. Kido, *Digital Fourier Analysis: Advanced Techniques*,
DOI: 10.1007/978-1-4939-1127-1_5,

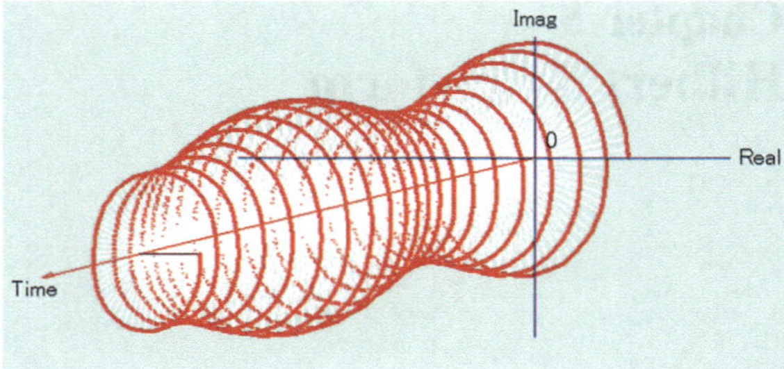

Fig. 5.1 Representation of a cosine wave with a slowly changing amplitude (*radius*). Animation available in supplementary files under filename E12-01_Envelop.exe

amplitude with time, once it is expressed in the form of Eq. (5.1), can be represented by a rotating vector whose length changes as a function of time.

The change of amplitude of a signal with time is actually the length change of the rotating vector that represents the signal with time. If the rate of change is slow, the time-dependent length change of the rotating vector can be referred to as the *envelope* of the waveform. By extending this idea, the length of the rotating vector, which is constant with $\sqrt{\cos^2(2\pi ft) + \sin^2(2\pi ft)} = 1$, is considered as the envelope of the cosine (or sine) wave.

By adopting the third axis that goes perpendicularly through the origin of the complex plane of Fig. 1.10 as the time axis, the rotating vector of a complex signal can be represented by the one as shown in Fig. 5.1.

We can call the length of the rotating vector as the envelope of a signal for a case when the amplitude changes slowly such as the one shown in Fig. 5.1. If the amplitude increases and decreases within one rotation, it cannot be called as the envelope. From Fig. 5.1, you will see that the envelope is the time-dependent radius of the cylinder which changes gradually as the time increases.

This understanding indicates that, in order to obtain the envelope of a wave with time-varying amplitude, it is necessary to have another wave that is orthogonal to the original wave. There are two waves that are orthogonal (i.e., different in phase by 90°) to the cosine wave $\cos(2\pi ft)$. If we take the real and imaginary parts of Euler's formula, the orthogonal wave to $\cos(2\pi ft)$ is

$$\cos(2\pi ft - \pi/2) = \sin(2\pi f)$$

The sine wave $\sin(2\pi ft)$ was directly made in the time domain by giving $\pi/2$ phase delay to the cosine wave $\cos(2\pi ft)$. However, we will adopt a method of finding orthogonal wave in the frequency domain. Once we find it, we can convert it into a time waveform by applying the inverse Fourier transform. If we have two orthogonal waves, the envelope of the original wave is given as the square root of the sum of squares of the two waves.

5.2 Orthogonal Wave to the Cosine Wave

If a wave is represented by the real part of a complex exponential function, the imaginary part is the orthogonal wave to the real part. That is, the real part of $\exp(j2\pi ft)$ is $\cos(2\pi ft)$ and the orthogonal wave to this is $\sin(2\pi ft)$, which is the imaginary part of $\exp(j2\pi ft)$. In general, if we can find a complex exponential function which has the given waveform as its real part, the orthogonal waveform is automatically given by its imaginary part. With this fact in mind, we come up to a method of obtaining exponential complex function that has the orthogonal wave to $\cos(2\pi ft + \varphi)$ as its real part, which is shown below.

① $\cos(2\pi ft + \varphi)$ is the real part of $\exp\{j(2\pi ft + \varphi)\}$. This is obvious from:

$$\exp\{j(2\pi ft + \varphi)\} = \cos(2\pi ft + \varphi) + j\sin(2\pi ft + \varphi)$$

② By multiplying $-j$ to $\exp(j2\pi ft + \varphi)$, the imaginary part becomes the real part.

$$-j\exp\{j(2\pi ft + \varphi)\} = -j\cos(2\pi ft + \varphi) + \sin(2\pi ft + \varphi) \quad (5.2)$$

The real part is the $\sin(2\pi ft + \varphi)$, which has the phase delay of $\pi/2$.

③ By multiplying $-j$ to Eq. (5.2),

$$-\exp\{j(2\pi ft + \varphi)\} = -\cos(j2\pi ft + \varphi) - j\sin(j2\pi ft + \varphi) \quad (5.3)$$

It is obvious that the real part of Eq. (5.3), $-\cos(2\pi ft + \varphi)$, is the $\pi/2$ delayed signal of $\sin(2\pi ft + \varphi)$.

④ By multiplying $-j$ to Eq. (5.3),

$$j\exp\{j(2\pi ft + \varphi)\} = j\cos(2\pi ft + \varphi) - \sin(2\pi ft + \varphi) \quad (5.4)$$

The real part of Eq. (5.4) is $-\sin(2\pi ft + \varphi)$, which has another $\pi/2$ phase delay.

⑤ By multiplying $-j$ to Eq. (5.4),

$$\exp\{j(2\pi ft + \varphi)\} = \cos(2\pi ft + \varphi) + j\sin(2\pi ft + \varphi) \quad (5.5)$$

This has the same form as Eq. (5.1) and its real part is the original waveform.

What was shown above is the use of the theory that multiplication by j is to advance the phase by $\pi/2$ and the division by j (or multiplication by $-j$) is to delay the phase by $\pi/2$ in the two-dimensional vector on the complex plane.

5.3 Discrete Cosine Waves and Orthogonal Waves

In order for the easiness of understanding and the direct connection to numerical operations, we will consider the N-point DFT of discrete cosine functions with the discrete frequency p $(0 \leq p \leq N)$.

(1) In the case of $x(n) = \cos(2\pi pn/N)$

$$X(k) = \mathrm{DFT}\{x(n)\} = \frac{1}{2}\sum_{n=0}^{N-1}\{\exp(j2\pi\frac{pn}{N}) + \exp(j2\pi\frac{pn}{N})\}\exp(-j2\pi\frac{kn}{N}) \qquad (5.6)$$
$$= \frac{N}{2}\{\delta(p-k) + \delta(p-N+k)\}$$

$$= \frac{N}{2}\{\delta(p-k) + \delta(p+k)\} \qquad (5.7)$$

where $\delta(m)$ is the unit impulse function with $\delta(m) = 1$ for $m = 0$ and $\delta(m) = 0$ for $m \neq 0$.

In Eq. (5.6), the range of k is from 0 to $N - 1$, but since the DFT is a circular sequence with period N, the range from $N/2$ to $N - 1$ corresponds to the negative frequency range. Therefore, the range of k in Eq. (5.7) is from $-N/2$ to $N/2 - 1$. Since the latter is easier in considering the relation with the continuous Fourier transform, Eq. (5.7) that has the direct representation of the negative frequency components will be used whenever appropriate. If it is necessary to consider the correspondence with the N-point DFT, you can add N to the negative frequency. The spectrum of Eq. (5.7) takes the value $N/2$ at $k = p$ and $-p$ and 0 at all other points. The spectrum of this function is shown by Fig. 5.2a.

How can we make an orthogonal wave to $\cos(2\pi pn/N)$? The multiplication of $-j$ does not give the desired function $\sin(2\pi pn/N)$. Since the two line spectra become complex and their signs are also changed, the positive frequency spectrum becomes that of $\sin(2\pi pn/N)$, but the negative frequency spectrum has the opposite sign with $\sin(2\pi pn/N)$. Instead of multiplying $-j$, the multiplication of $-j\,\mathrm{sgn}(k)$ to the spectrum of $\cos(2\pi pn/N)$ gives what we want, where $\mathrm{sgn}(k)$ indicates the sign of k.

The multiplication of $-j\,\mathrm{sgn}(k)$ is to multiply $-j$ when $k > 0$, and to multiply $+j$ when $k < 0$. Then we have

$$-j\,\mathrm{sgn}(k)\mathrm{DFT}\left[\cos\left(2\pi\frac{pn}{N}\right)\right] = -j\,\mathrm{sgn}(k)\frac{N}{2}\{\delta(p-k) + \delta(p+k)\}$$
$$= j\frac{N}{2}\{-\delta(p-k) + \delta(p+k)\} \qquad (5.8)$$

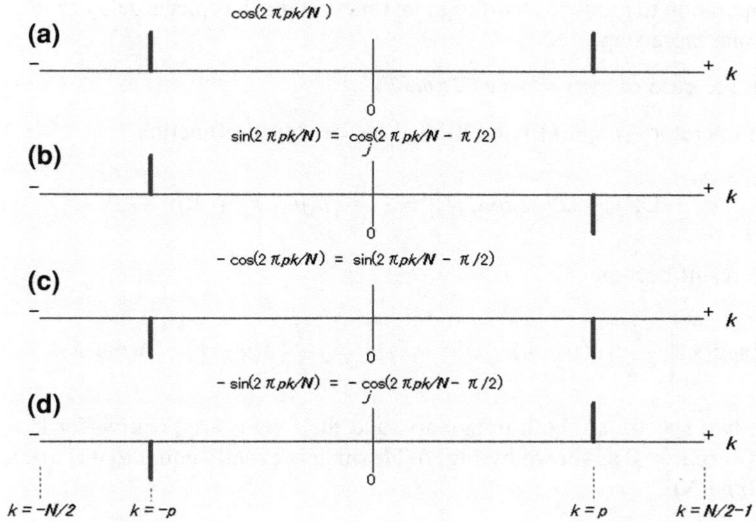

Fig. 5.2 Four orthogonal functions with phase delays in π/2 steps starting from cos (2πpn/N)

Now both spectra are imaginary and the signs of the spectra are negative for $k > 0$ and positive for $k < 0$, which are exactly the spectra of $\sin(2\pi pn/N)$. This is shown by Fig. 5.2b.

By multiplying $-j\,\mathrm{sgn}(k)$ to the spectrum of $\cos(2\pi pn/N)$, the orthogonal function $\sin(2\pi pn/N)$, which has π/2 phase delay from $\cos(2\pi pn/N)$ has been obtained. Can we do the same thing to obtain an orthogonal function with phase delay π/2?

(2) In the case of $x(n) = \sin(2\pi pn/N)$

We will start from

$$\mathrm{DFT}\left[\sin\left(2\pi\frac{pn}{N}\right)\right] = j\frac{N}{2}\{-\delta(p-k)+\delta(p+k)\} \tag{5.9}$$

By multiplying $-j\,\mathrm{sgn}(k)$ to Eq. (5.9) (Fig. 5.2b), both line spectra become real and both have negative signs as shown by Eq. (5.10).

$$-j\,\mathrm{sgn}(k)\left[j\frac{N}{2}\{-\delta(p-k)+\delta(p+k)\}\right] = -\frac{N}{2}\{\delta(p-k)+\delta(p+k)\} \tag{5.10}$$

This is the spectrum shown in Fig. 5.2c. Compared to (a), both spectra has opposite signs showing that this is equal to the spectrum of $-\cos(2\pi pn/N)$. The fact that the function that has phase delay of π/2 from $\sin(2\pi pn/N)$ is $-\cos(2\pi pn/N)$ is shown by Eqs. (5.2) and (5.3). It was shown that multiplication of $-j\,\mathrm{sgn}(k)$ works as an operation to produce an orthogonal function that has phase delay of $\pi/2$. Let us check one more step.

(3) In the case of $x(n) = -\cos(2\pi pn/N)$

The operator $-j$ sgn(k) is multiplied to the starting function.

$$\mathrm{DFT}[-\cos(2\pi pn/N)] = -\frac{N}{2}\{\delta(p-k) + \delta(p+k)\} \tag{5.11}$$

The result becomes

$$-j\,\mathrm{sgn}(k)\left[-\frac{N}{2}\{\delta(p-k) + \delta(p+k)\}\right] = j\frac{N}{2}\{\delta(p-k) - \delta(p+k)\} \tag{5.12}$$

The two spectra are both imaginary, and their signs are positive for k > 0 and negative for k < 0 as shown by Fig. 5.2d which is exactly equal to the spectrum of $-\sin(2\pi pn/N)$.

(4) In the case of $x(n) = -\sin(2\pi pn/N)$

By multiplying $-j$ sgn(k) to

$$\mathrm{DFT}[-\sin(2\pi pn/N)] = j\frac{N}{2}\{\delta(p-k) - \delta(p+k)\} \tag{5.13}$$

we have

$$-j\,\mathrm{sgn}(k)\left[j\frac{N}{2}\{\delta(p-k) - \delta(p+k)\}\right] = \frac{N}{2}\{\delta(p-k) + \delta(p+k)\} \tag{5.14}$$

The two spectra are both real, and their signs are both positive for the positive and negative frequencies. We now came back to the original function shown by Fig. 5.2a.

Equations (5.2)–(5.5) show the functions starting from cosine functions and followed by functions with phase delays in $\pi/2$ steps. Equations (5.7)–(5.14) and Fig. 5.2 show those in frequency domains.

The above discussions showed that the multiplication of $-j$ sgn(k) to the spectrum of a signal gives the phase delay of $\pi/2$. It is obvious that the multiplication of j sgn(k) gives the phase advance of $\pi/2$ to the signal. The understanding that the multiplication of j gives the phase advance of $\pi/2$ is always correct in the sense that the phase is increased by $\pi/2$ in the counter-clockwise rotation. However, for a signal with a negative frequency, i.e., for a clockwise rotating vector, the counter-clockwise increase of $\pi/2$ is equivalent to rotating backward or going back in time. For this reason, $-j$ must be multiplied to a signal with a negative frequency to increase the clockwise rotating angle. Therefore, for a signal with positive and negative frequency components, the multiplication of $-j$ sgn(k) is necessary to produce an orthogonal signal with the phase delay of $\pi/2$.

When we apply the multiplier (or operator) $-j$ sgn(k) to a signal, we assume that the range of k is $-N/2 \leq k \leq N/2$. This cannot be directly applied to the case of DFT, in which the range of k is $0 \leq k < N$. In this case, you can use

$$-j\,\mathrm{sgn}(N/2-k) = \begin{cases} -j & (0 \leq k < N/2) \\ +j & (N/2 \leq k < N) \end{cases}$$

in order to produce a signal with the phase delay of $\pi/2$.

Our goal is to produce orthogonal signals to signals with arbitrary waveforms other than the simple sine or cosine waves as discussed above. From the knowledge of the Fourier transform, we may be able to say that, if frequency components of one signal are all delayed by $\pi/2$ to those of another signal, the former is orthogonal to the latter signal. If we define the "orthogonality" of signals in this way, we can produce orthogonal signals by multiplying $-j$ sgn($N/2 - k$) to the spectrum obtained by the DFT of arbitrary signal and then applying IDFT to the modified spectrum. We will check it in the next section.

5.4 Generation of Orthogonal Waveforms

We considered the cases of sine and cosine functions in Sect. 5.3. In this section, we will deal with arbitrary signals. The Fourier series analysis teaches us that every signal can be expressed as a summation of sine and cosine functions, all those components are orthogonal with each other (see Sect. 2.1 of vol. I). Also the Fourier transform is linear. Therefore, it is natural to consider that, by applying the method described in the preceding section to all components of the Fourier series, an orthogonal signal can be obtained.

Let us consider generating an orthogonal waveform to a sample sequence $x(n)$. The sequence $x(n)$ can be expressed using its DFT $X(k) = R(k) + jI(k)$ as (see Appendix 5A)

$$x(n) = \frac{1}{N}\sum_{k=0}^{N-1}\left\{R(k)\cos(2\pi\frac{kn}{N}) - I(k)\sin(2\pi\frac{kn}{N})\right\} \tag{5.15}$$

where n is the discrete time ($n = 0, 1, 2, \ldots, N - 1$) and k is the discrete frequency ($k = 0, 1, 2, \ldots, N - 1$).

Once the signal is expressed in this way in the time domain, the orthogonal signal can be obtained by adding $-\pi/2$ for $0 \leq k < N/2$ and $\pi/2$ for $N/2 \leq k < N$ to each of sine and cosine components. This is a direct method but it needs to deal with individual frequency components. Also it is not using the method in the frequency domain derived in the preceding section.

In order to take the frequency domain method, we will start from the Fourier expansion equation using the complex exponential functions, i.e., the N-point IDFT of $X(k)$.

$$x(n) = \text{IDFT}[X(k)] = \frac{1}{N}\sum_{k=0}^{N-1} X(k)\exp(j2\pi\frac{kn}{N}) \tag{5.16}$$

The range of k in the above equation is from 0 to $N - 1$, but it is more convenient to change the range from $-N/2$ to $N/2 - 1$ in order for the operation to multiply $-j$ sgn(k). The result is not changed due to the circularity of the DFT as explained before. The equation used is slightly changed from Eq. (5.16).

$$x(n) = \text{IDFT}[X(k)] = \frac{1}{N}\sum_{k=-N/2}^{N/2-1} X(k)\exp(j2\pi\frac{kn}{N}) \tag{5.17}$$

where $X(-k) = X(N - k) = X^*(k)$.

If we use the method introduced in the preceding section, the orthogonal spectrum to the spectrum $X(k)$ is obtained by multiplying $-j$ sgn(k) to the spectrum $X(k) = R(k) + jI(k)$ of Eq. (5.17). The direct current component must be removed since the DC has no phase and there is no orthogonal component to it. Therefore, we deal with waveforms with no DC component (i.e., $R(0) = I(0) = 0$), hereafter.

The integer N can be either odd or even. Since the generality is not lost, however, we assume that N is even. Furthermore, we assume that $X(-N/2) = 0$ (i.e., $R(-N/2) = I(-N/2) = 0$). With these assumptions in mind, we will step forward in handling some more equations.

The orthogonal spectrum $X_\perp(k)$ of $X(k)$ is given by Eq. (5.18).

$$\begin{aligned} X_\perp(k) &= -j\,\text{sgn}(k)X(k) = -j\,\text{sgn}(k)[R(k) + jI(k)] \\ &= \text{sgn}(k)I(k) - j\,\text{sgn}(k)R(k) \end{aligned} \tag{5.18}$$

The orthogonal sequence $x_\perp(n)$ to the original sequence $x(n)$ is given by the IDFT of the spectrum given by Eq. (5.18).

$$x_\perp(n) = \frac{1}{N}\sum_{k=-N/2}^{N/2-1} \{\text{sgn}(k)I(k) - j\,\text{sgn}(k)R(k)\}\exp(j2\pi\frac{kn}{N}) \tag{5.19}$$

If this is orthogonal to $x(n)$, the following relation must hold.

$$\sum_{n=0}^{N-1} x(n)x_\perp(n) = 0 \tag{5.20}$$

Since it is too lengthy to show here if this is true or not, the proof is given in Appendix 5B.

5.5 Hilbert Transform

In Sect. 5.4, equations of discrete sequences by the DFT have been shown in order to generate orthogonal sequences in the discrete frequency domain. The same thing must hold for the continuous signals and their spectra.

Following the method we derived in the preceding section, if $X(f)$ is the spectrum of $x(t)$, the orthogonal spectrum (delayed and advanced by $\pi/2$ in the positive and negative frequency domains, respectively) is generated by multiplying $-j$ sgn(k) to $X(f)$.

$$X_{\perp}(f) = -j\,\mathrm{sgn}(f)X(f) \tag{5.21}$$

Then, the inverse Fourier transform of $X_{\perp}(f)$ is performed. Since the results is inverse Fourier transform of the orthogonal spectrum to the spectrum $X(f)$ of $x(t)$, it must be the orthogonal signal to $x(t)$.

$$x_{\perp}(t) = -j\int_{-\infty}^{+\infty}\mathrm{sgn}(f)X(f)\exp(j2\pi ft)df \tag{5.22}$$

Up to here, the process is the same as in the preceding section. Now we will derive an equation that gives $X_{\perp}(f)$ from $x(t)$, which is the same as Eq. (5.22) but in a different form. In Chap. 1 Eq. (1.8), we have learned that the inverse transform of two frequency domain functions is given by the convolution of their respective time domain functions.

Therefore, Eq. (5.22) is given as the convolution of the inverse Fourier transforms of $-j$ sgn(k) and $X(f)$. The inverse Fourier transform of $-j$ sgn(k) is given by

$$F^{-1}[-j\,\mathrm{sgn}(f)] = \frac{1}{\pi t} \tag{5.23}$$

as shown in Appendix 5C. Then, Eq. (5.22) is given as the convolution of $1/(\pi t)$ and $x(t)$.

$$x_{\perp}(t) = \frac{1}{\pi}\int_{-\infty}^{\infty}\frac{1}{t-\tau}x(\tau)d\tau \tag{5.24}$$

This is known as the *Hilbert transform*. The integral may have the divergence problem but it is avoided by interpreting the integral as

$$x_{\perp}(t) = \lim_{\Delta\to 0}\left[\frac{1}{\pi}\int_{-\infty}^{t-\Delta}\frac{1}{t-\tau}x(\tau)d\tau + \frac{1}{\pi}\int_{t+\Delta}^{\infty}\frac{1}{t-\tau}x(\tau)d\tau\right]$$

Now, let us calculate Eq. (5.22) with $x(t) = \cos(2\pi f_0 t)$.

$$x_{\perp}(t) = \frac{1}{\pi}\int_{-\infty}^{\infty}\frac{\cos(2\pi f_0 \tau)}{t-\tau}d\tau$$

By introducing a new parameter $z = t - \tau$, the above equation is rewritten as

$$\begin{aligned}
x_{\perp}(t) &= \frac{1}{\pi}\int_{-\infty}^{\infty}\frac{\cos\{2\pi f_0 t - 2\pi f_0 z)}{z}dz \\
&= \frac{1}{\pi}\int_{-\infty}^{\infty}\frac{\cos(2\pi f_0 t)\cos(2\pi f_0 z) + \sin(2\pi f_0 t)\sin(2\pi f_0 z)}{z}dz \\
&= \frac{\cos(2\pi f_0 t)}{\pi}\int_{-\infty}^{\infty}\frac{\cos(2\pi f_0 z)}{z}dz + \frac{\sin(2\pi f_0 t)}{\pi}\int_{-\infty}^{\infty}\frac{\sin(2\pi f_0 z)}{z}dz
\end{aligned}$$

Since the integrand of the first term is odd, it is zero. The second term is given by

$$\begin{aligned}
x_{\perp}(t) &= 2\frac{\sin(2\pi f_0 t)}{\pi}\int_{0}^{\infty}\frac{\sin(2\pi f_0 z)}{z}dz \\
&= \frac{2\sin(2\pi f_0 t)}{\pi}\frac{\pi}{2} \\
&= \sin(2\pi f_0 t).
\end{aligned}$$

This calculation is in the time domain. If we utilize the fact that Eq. (5.24) is the convolution of $1/(\pi t)$ and $\cos(2\pi f_0 t)$, the same result is obtained in the frequency domain by applying the inverse Fourier transform to the product of the Fourier transforms of $1/(\pi t)$ and $\cos(2\pi f_0 t)$.

First, the Fourier transform of $1/(\pi t)$ is given by $-j\ \mathrm{sgn}(f)$ from Eq. (5.23) and the Fourier transform of $\cos(2\pi f_0 t)$ is given by (see Appendix 2B)

$$F\{\cos(2\pi f_0 t)\} = \frac{1}{2}[\delta(f + f_0) + \delta(f - f_0)] \tag{5.25}$$

Their product is then given by

$$\frac{1}{2}[j\delta(f + f_0) - j\delta(f - f_0)]$$

This is equal to the spectrum of $\sin(2\pi f_0 t)$ as shown in Appendix 2B. The orthogonal function $x_{\perp}(t)$ of $\cos(2\pi f_0 t)$ is given by the inverse Fourier transform of the above equation as follows:

$$x_{\perp}(t) = \sin(2\pi f_0 t).$$

The Hilbert transform of a function gives the phase delay of π/2 to its positive frequency components and the phase advance of π/2 to its negative frequency components. Therefore, if we multiply *j* sgn(*f*) to the Fourier transform of $x_{\perp}(t)$ and then apply the inverse Fourier transform to it, the original function *x*(*t*) is obtained. By applying the same procedure from Eqs. (5.21)–(5.24), the inverse Hilbert transform is given by

$$x(t) = -\frac{1}{\pi}\int_{-\infty}^{\infty}\frac{1}{t-\tau}x_{\perp}(\tau)d\tau \tag{5.26}$$

The inverse Hilbert transform is the same transform with the (forward) Hilbert transform except for the sign reversal.

5.6 Real and Imaginary Parts of Transfer Functions

The transfer function *H*(*f*) of a transfer system with the impulse response *h*(*t*) is given by the Fourier transform of *h*(*t*). The physically realizable systems satisfy the causality; i.e., there is no response before the input. The impulse response *h*(*t*) of causal systems must be zero for $t < 0$ (if $t = 0$ is the time of input). The Fourier transform of a function that satisfies the causality has the property that the real and imaginary parts are mutually given by the Hilbert and inverse Hilbert transforms.

As the left side of Fig. 5.3 shows, the impulse response *h*(*t*) can be represented by the summation of even and odd functions, $h_E(t)$ and $h_O(t)$.

$$h(t) = h_E(t) + h_O(t) \tag{5.27}$$

The even function $h_E(t)$ and the odd function $h_O(t)$ are equal to *h*(*t*)/2 for $t > 0$ and their signs are reversed for $t < 0$. Therefore, $h(t) = 0$ for $t < 0$ is automatically satisfied and there is no need to restrict the range within $t \geq 0$.

Since the inverse Fourier transforms of the real and imaginary parts of the Fourier transform of *h*(*t*) are the even function $h_E(t)$ and odd function $h_O(t)$, respectively, the Fourier transform *H*(*f*) is represented by Eq. (5.28) using their Fourier transforms, $H_E(f)$ and $H_O(f)$.

$$\begin{aligned} H(f) &= \int_{-\infty}^{+\infty} h(t)\exp(-j2\pi ft)dt \\ &= \int_{-\infty}^{+\infty} h_E(t)\exp(-j2\pi ft)dt + \int_{-\infty}^{+\infty} h_O(t)\exp(-j2\pi ft)dt \\ &= H_E(f) + H_O(f) \end{aligned} \tag{5.28}$$

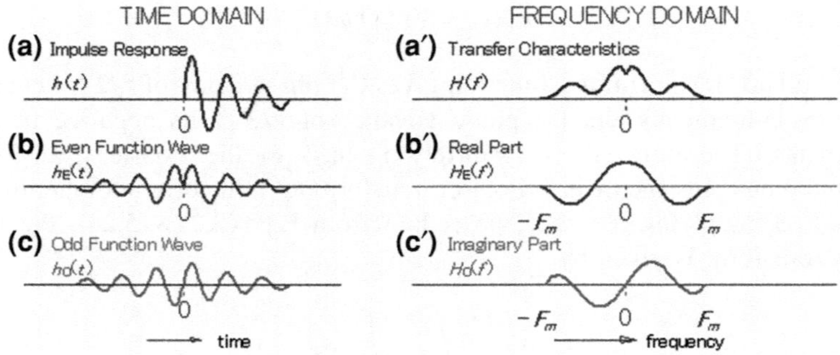

Fig. 5.3 Representation of an impulse function by even and odd functions and their Fourier transforms

Appendix 2C shows that the Fourier transform of $H_E(f)$ with respect to frequency f is equal to $h_E(-t)$, which is equal to $h_E(t)$ since it is an even function. By multiplying sgn(t) to it in order to reverse the sign in the negative time region, it becomes $h_O(t)$.

$$h_O(t) = \mathrm{sgn}(t)h_E(t) \tag{5.29}$$

Then, $H_O(f)$, the Fourier transform of $h_O(t)$, is given as the Fourier transform of $\mathrm{sgn}(t)h_E(t)$.

$$H_O(f) = \int_{-\infty}^{+\infty} \mathrm{sgn}(t)h_E(t)\exp(-j2\pi ft)dt$$

This equation shows that $H_O(f)$ is the Fourier transform of the product of the two time functions. Then Eq. (5.30) is derived by the same way as Eq. (5.24) was obtained from Eq. (5.22).

$$H_O(f) = \frac{1}{\pi}\int_{-\infty}^{+\infty} \frac{1}{f-\varphi}H_E(\varphi)d\varphi \tag{5.30}$$

Equation (5.30) shows that the relationship of the Hilbert transform exists between the real and imaginary parts of the spectrum of a causal impulse response.

5.7 Envelopes and Orthogonal Functions

One purpose of generating orthogonal functions was to obtain envelopes of functions using the property that the squared sum of an original and its orthogonal functions gives the envelope of the original function (see Fig. 5.1). The simplest

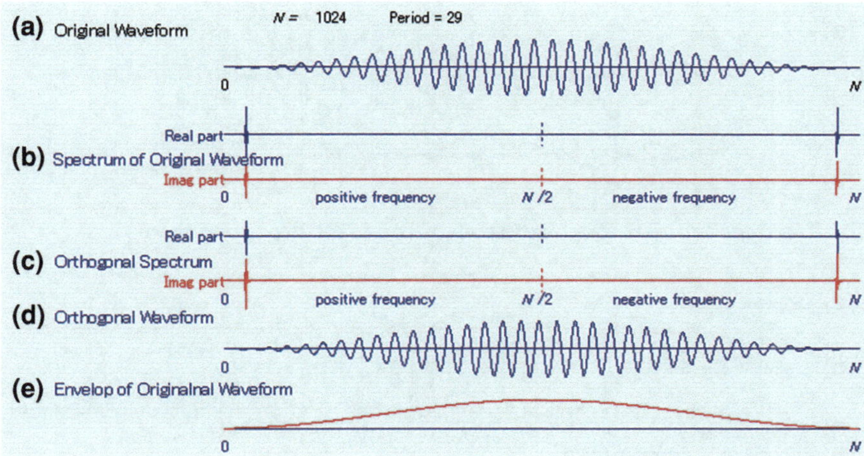

Fig. 5.4 Generation of an envelope function of a 1,024-point Hanning windowed sine wave with 29 periods in the window. **a** Hanning-windowed waveform, **b** DFT of (**a**), **c** orthogonal function of (**b**), **d** IDFT of (**c**), and **e** envelop of (**a**) that is the square root of the squared sum of (**a**) and (**d**). Animation available in supplementary files under filename E12-04_Envelop.exe

example is the two orthogonal functions, $\cos(2\pi ft + \varphi)$ and $\sin(2\pi ft + \varphi)$, which have the constant envelope (square root of the squared sum is 1). We will check this using more general waveforms.

First, we will check that the square root of the squared sum of the sine wave and its orthogonal wave produced using Eq. (5.9) becomes the envelope of the sine wave. Figure 5.4 shows the process. Figure 5.4a is the waveform of the 1,024-point Hanning-windowed sine wave with period length of 29. Since there are non-integer number of periods (=frequency) within the window, its spectrum has the real and imaginary parts. Figure 5.4b shows the spectrum. The orthogonal spectrum, Fig. 5.4c, is obtained by multiplying $-j\ \mathrm{sgn}(N/2 - k)$ to (b). This is accomplished by reversing only the sign of the real part in the positive frequency region ($0 < k < N/2$) of (b) and then substituting it into the imaginary part of (c) and by reversing only the sign of the imaginary part in the negative frequency region ($N/2 < k < N$) of (b) and substituting it into the real part of (c). Figure 5.4d is the inverse Fourier transform of (c) and the square root of the squared sum of (a) and (d) is (e), which is the shape of the Hanning window. No one may have an objection to say that (e) is the envelope of (a).

Let us see if the original waveform contains the second harmonics. Figure 5.5 shows the results. In this case the squared sum of the original and orthogonal functions, Fig. 5.5e, does not give the Hanning window function. It is wavy in the whole region indicating that the square root of the squared sum of the original and orthogonal functions does not necessarily give the envelope of the original function.

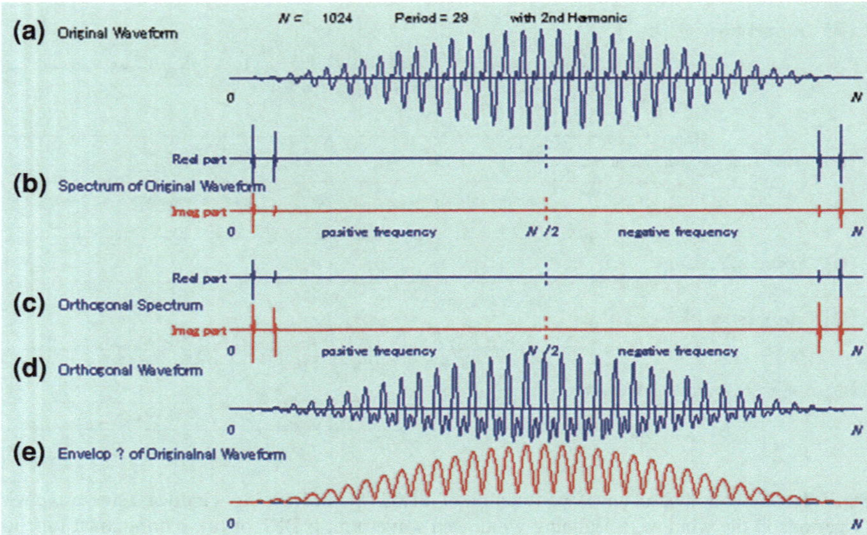

Fig. 5.5 Generation of an envelope function of a 1,024-point Hanning-windowed sine waves with 29 and 14.5 periods in the window (the latter is the second harmonic of the former), **a** Hanning-windowed waveform, **b** DFT of (**a**), **c** orthogonal function of (**b**), **d** IDFT of (**c**), **e** square root of the squared sum of (**a**) and (**d**). Animation available in supplementary files under filename E12-05_Envelop.exe

The reason that the desired envelope shape is not obtained when the original waveform contains two sine waves is that the length of the rotating vector that represent the waveform changes during one rotation.

Another example is shown in Fig. 5.6 where the period of the second sine component is changed to 33 from 58 in Fig. 5.5. In this case, the perfect envelope is obtained. Since the waveform is the sum of two sine waves with the same amplitude and the frequency difference of 4, it takes the maximum amplitude when difference of the instantaneous phases of the two waves is zero and it becomes zero when difference of the instantaneous phases of the two waves is π. Therefore, the envelope waveform is the full-wave rectified waveform of the sine wave with half of the frequency difference of the two waves.

In the program attached to Fig. 5.6, you can change the frequency and amplitude of the second wave. Results will give you idea on how the envelopes of waveforms with two sine waves are determined.

The difference between Figs. 5.5 and 5.6 is frequency separation between the two components. In Fig. 5.5, the discrete frequencies are 35.3 (=1,024/29) and 70.6 (=35.3 × 2), whereas in Fig. 5.6, those are 35.3 and 31.03 (=1,024/33). If the separation is large, the variation of the waveform becomes complex and the envelope obtained by the present method does not really represent the envelope of the waveform.

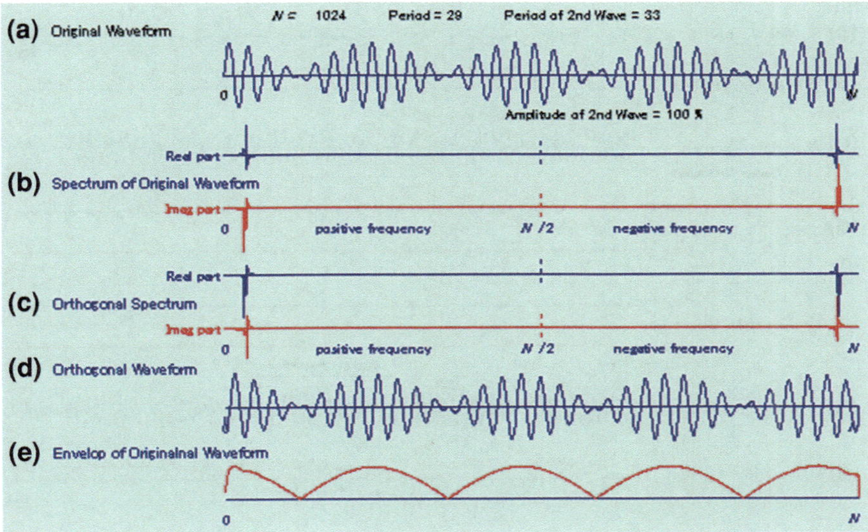

Fig. 5.6 Generation of an envelope function of a 1,024-point Hanning-windowed sine waves with periods of 29 and 33, **a** Hanning-windowed waveform, **b** DFT of (**a**), **c** orthogonal function of (**b**), **d** IDFT of (**c**), and **e** square root of squared sum of (**a**) and (**d**). Animation available in supplementary files under filename E12-06_Envelop.exe

Let us consider briefly the spectrum of the time-windowed sine function. Its spectrum has the shape such that the line spectrum of the sine function is surrounded by the spectrum of the window function. This is the same with the spectrum of an amplitude modulated sine function. Around the line spectrum of the sine wave, the spectrum of the amplitude modulating function is distributed.

Envelopes are well defined only when a period of the amplitude modulating function is much longer than the period of the sine function. In such a case, the amplitude of the sine function changes slowly, and the spectrum spreading is narrow. Amplitude-modulated signals with such spectrum will be discussed in the next section.

5.8 Amplitude Modulation

The amplitude-modulated wave is a sine (including cosine) wave, whose amplitude is modulated so that its envelope becomes the signal waveform. The sine wave is called *carrier*, and the signal that modifies the amplitude of the carrier is called *modulator*. Modifying the carrier amplitude by the modulator is called *amplitude modulation* (abbreviated as *AM*).

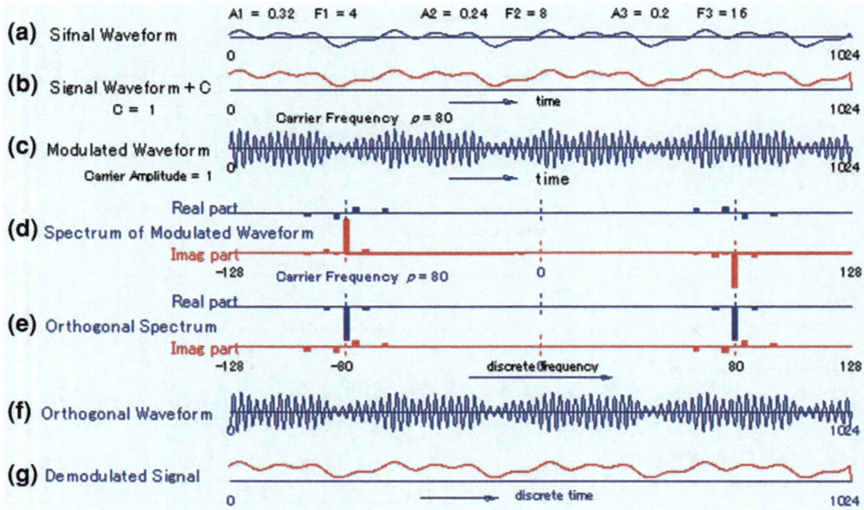

Fig. 5.7 Amplitude modulation and demodulation by the envelope detection. Animation available in supplementary files under filename E12-07_AM.exe

The relation between the carrier and the modulator is shown in Fig. 5.7. Figure 5.7a is the signal wave and (b) is a modulator made of (a) by adding DC so that it never becomes negative. The product of (b) and a carrier, which has a much higher frequency, p, than (b), is shown by (c).

Let us consider the frequency components of the amplitude modulated wave, Fig. 5.7c. Let the carrier be $\sin(2\pi pn/N)$ and the signal wave be $x(n)(|x(n)| \leq 1)$. Then, Fig. 5.7c is given by the next equation.

$$\begin{aligned} w(n) &= \{1 + x(n)\} \sin(2\pi \frac{pn}{N}) \\ &= \sin(2\pi \frac{pn}{N}) + x(n) \sin(2\pi \frac{pn}{N}) \end{aligned} \tag{5.31}$$

The spectrum of this discrete waveform is calculated in the following way.

The DFT of the first term is given by Eq. (5.9). The DFT of the second term is the convolution of Eq. (5.9) and $X(k)$, where $X(k)$ is the DFT of $x(n)$. Then, it is given by

$$
\begin{aligned}
\mathrm{DFT}\left[x(n)\sin(2\pi\frac{pn}{N})\right] &= \sum_{n=0}^{N-1} x(n)\sin(2\pi\frac{pn}{N})\exp(-j2\pi\frac{kn}{N}) \\
&= \frac{1}{j2}\sum_{n=0}^{N-1} x(n)\left\{\exp(j2\pi\frac{pn}{N}) - \exp(-j2\pi\frac{pn}{N})\right\}\exp(-j2\pi\frac{kn}{N}) \\
&= \frac{1}{j2}\sum_{n=0}^{N-1} x(n)\left\{\exp(-j2\pi\frac{(k-p)n}{N}) - \exp(-j2\pi\frac{(k+p)n}{N})\right\} \\
&= \frac{1}{j2}\{X(k-p) - X(k+p)\}
\end{aligned}
$$

Therefore, the DFT of $w(n)$ is given by the next equation Eq. (5.32).

$$
W(k) = \frac{1}{j2}\{-N\delta(k-p) + N\delta(k+p) + X(k-p) - X(k+p)\} \tag{5.32}
$$

Figure 5.7d shows the spectrum given by Eq. (5.32). The first two terms are shown by the tall vertical bars and the last two terms are shown by the short bars around the tall bars. The orthogonal spectrum is given by multiplying $-j$ sgn(k) to (d), which is shown by (e). The IDFT of (e) is shown by (f), which is the orthogonal waveform to (c). Figure 5.8g is the square root of the sum of squares of (c) and (f), which is equal to (b).

The spectra of Fig. 5.7c, d show that the spectra of $X(k)$ are distributed around the frequency $\pm p$ of the carrier. Therefore, it seems that the modulator (a) is recovered from the modulated waveform (c) if the range of the spectrum of $X(k)$ is narrower than p. Otherwise, the spectral distributions centered at $+p$ and $-p$ will overlap with each other by crossing the border ($k = 0$), and the recovery is impossible. However, this is only one necessary condition. We have already seen this example in Fig. 5.5.

The above problem is due to the frequency range of the modulating signal. Another problem is caused if the amplitude of the modulating signal is too large. This example is shown in Fig. 5.8. In this example, the amplitude of the modulating signal is larger than C, the magnitude of DC. In such a case, there is regions in which (b) becomes negative. In those regions, the phase of the carrier is reversed. The demodulated signal (g) obtained by the same procedure as before is not the same with the modulating signal (a), i.e., the recovery is impossible. This problem is of course avoided by letting C larger than the modulating signal.

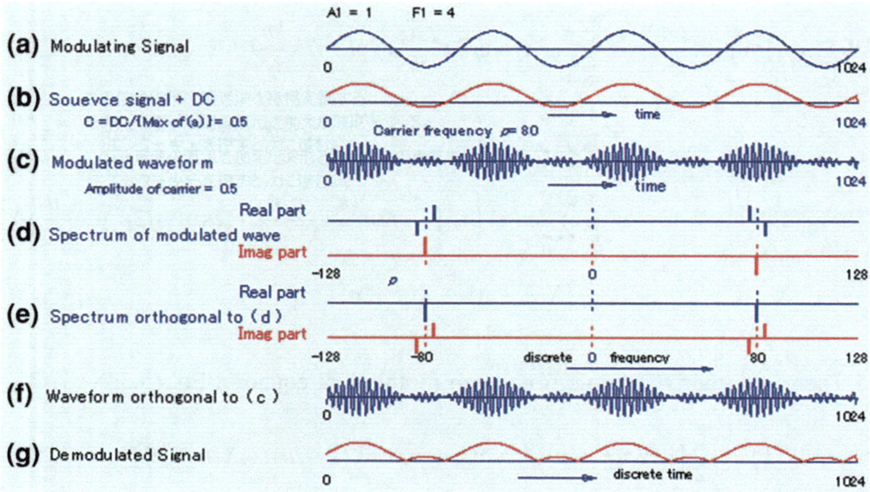

Fig. 5.8 Example of amplitude modulation and demodulation when the modulating signal is too large. Animation available in supplementary files under filename E12-08_AM.exe

5.9 Instantaneous Frequency

We learned how to make the waveform which is orthogonal to a given waveform. Also we learned that the complex signal with the latter as its real part and the former as its imaginary part is a rotating vector on the complex plane. Generally, the length and the rotation speed of the vector changes in time. The rotation angle at an instance is the *instantaneous phase* and its time derivative, i.e., the angle velocity is the *angular frequency*. The *instantaneous frequency* is given by dividing the angular frequency by 2π, i.e., the number of rotation per second.

If we have an original waveform $x(t)$ and its orthogonal waveform $x_{\perp}(t)$, we can obtain the instantaneous phase and instantaneous frequency. The step is as follows.

① Derive $x_{\perp}(t)$ from $x(t)$.

② Calculate the instantaneous phase $\theta(t)$ as follows:

$$\theta(t) = \tan^{-1}[x_{\perp}(t)/x(t)] \tag{5.33}$$

③ Differentiate $\theta(t)$ with respect to t to obtain the angular frequency $\omega(t)$.

$$\omega(t) = \frac{\partial\theta(t)}{\partial t}$$

④ Divide by 2π to obtain the instantaneous frequency

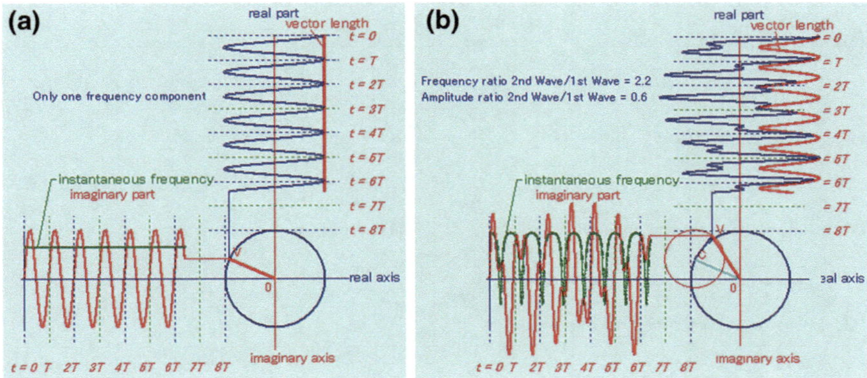

Fig. 5.9 Rotating vector of the signal given by Eq. (5.32) on the complex plane. The angular velocity of the vector $\overrightarrow{OV}$ is the instantaneous angular frequency. Animation available in supplementary files under filenames E12-09a_1CompoV.exe and E12-09b_2CompoV.exe

$$f(t) = \frac{\omega(t)}{2\pi} = \frac{1}{2\pi}\frac{\partial\theta(t)}{\partial t}$$

An application example of the above process to a single-sine signal and a double-sine signal is shown in Fig. 5.9a, b, respectively. The equation is given by

$$x(t) = A_1 \cos(2\pi f_1 t) + A_2 \cos(2\pi f_2 t) \tag{5.35}$$

where the parameters are $A_2 = 0$ for a single-sine wave, and $A_2/A_1 = 0.4$ and $f_2/f_1 = 2.2$ for a double-sine wave. The rotation vectors for the single and double-sine signals are shown in Fig. 5.9a, b, respectively.

In the case of the single-sine signal, the length and the rotation speed of the rotation vector $\overrightarrow{OV}$ are constant. However, in the case of the double-sine signal, the rotation of the added vector is not simple since the origin of the vector for the second sine is the top of the vector of the first sine. The first sine is represented by the rotating vector $\overrightarrow{OC}$that rotates around the origin of the complex plane with the angular velocity $2\pi f_1$ and the constant amplitude A_1. Then, the second sine is represented by the rotating vector $\overrightarrow{CV}$that rotates around point C, the top of the vector $\overrightarrow{OC}$, with the angular velocity $2\pi f_2$ and the constant amplitude A_2. The real signal $x(t)$ given by Eq. (5.35) is the projection of the vector $\overline{0V}$ (=$\overrightarrow{OC} + \overrightarrow{CV}$) onto the real axis, which is shown by the curve going downward along the imaginary axis. On the other hand, the orthogonal signal $x_\perp(t)$ is the projection of the vector $\overline{0V}$ onto the imaginary axis, which is shown by the curve going rightward along the real axis.

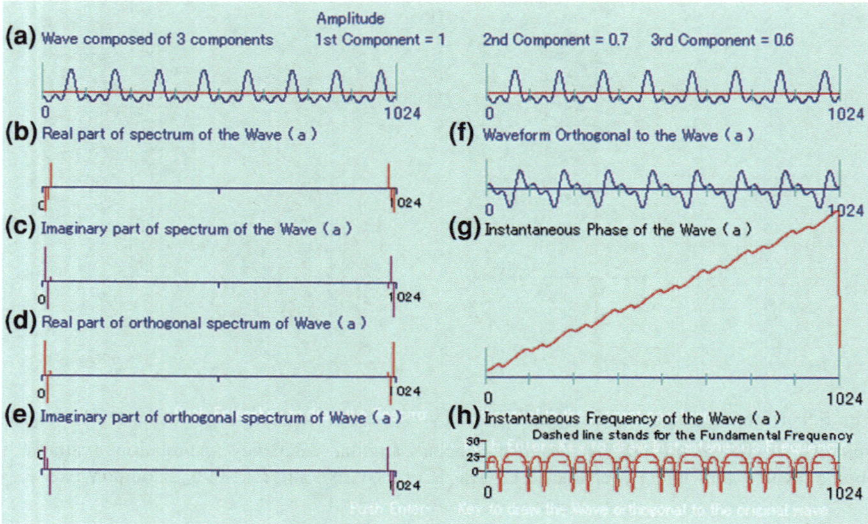

Fig. 5.10 Signal with three sine components and its instantaneous frequency. **a** Signal waveform, **b** and **c** real and imaginary parts of the spectrum of (**a**), **d** and **e** real and imaginary parts of the orthogonal spectrum of (**b**) and (**c**), **f** orthogonal signal of (**a**), and **g** instantaneous frequency. Animation available in supplementary files under filename E12-10_InstF_A.exe

The rotating vector $\overrightarrow{OC}$ rotating with the constant speed is the same with the vector $\overline{0V}$ in the case of the single-sine signal. In the case of the double-sine signal, the increment rate of the vector $\overline{0V}$ speeds up or slows down depending on the direction of the second vector $\overrightarrow{CV}$, which is represented by the instantaneous frequency (thick line) shown together with $x_{\perp}(t)$ along the horizontal axis. In this example, the instantaneous frequency becomes negative two times per one revolution of the vector $\overrightarrow{OC}$.

Figure 5.10 shows an example of a signal with three sine components with $A_1 = 1$, $f_1 = 8$, $A_1 = 0.8$, $f_2 = 16$, and $A_3 = 0.4$, $f_3 = 32$ ($N = 1{,}024$). The operator $-j\ \mathrm{sgn}\{X(f)\}$ is multiplied to the spectra (b) and (c) of the waveform (a) and the orthogonal spectra (d) and (e) are obtained and then (f) is calculated by the IDFT. From (a) and (f), the instantaneous phase (g) and its time derivative, the instantaneous frequency (h) are obtained. In this example, the instantaneous frequency periodically takes negative values because the amplitudes of the second and fourth harmonics are relatively large compared to the fundamental.

As shown above, the instantaneous frequency of a signal composed of sine waves with large frequency separation is largely dependent on time. For example, if the same procedure is applied to the vowel /a/ shown by the thin line in Fig. 5.11a, the orthogonal waveform takes the waveform shown by (b), and the instantaneous frequency shown by (c) largely fluctuates. Since the instantaneous

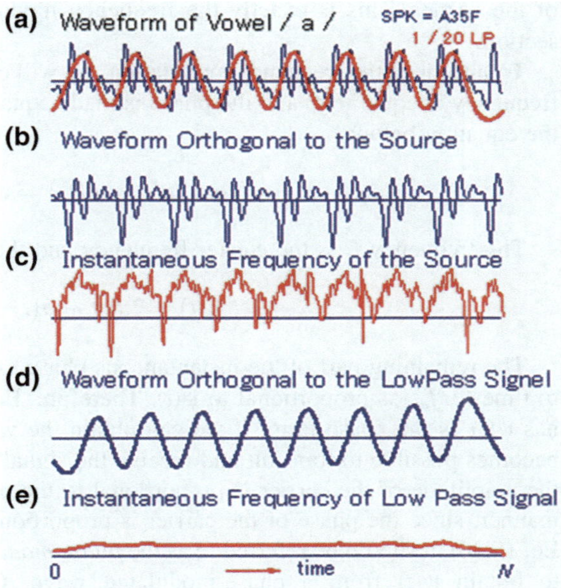

Fig. 5.11 Instantaneous frequencies of a vowel /a/. Animation available in supplementary files under filename E12-11_InstF_V.exe

frequency is affected by many frequency components, it cannot be directly used for the pitch detection of a speech.

If the higher frequency components are depressed by a low-pass filter, the waveform becomes like the one shown by the thick line in Fig. 5.11a. The orthogonal waveform of (a) is (d), and the instantaneous frequency given by the low-pass filtered waveform of (a) and (d) is (e). In this case, the instantaneous frequency is smooth and seems to show the pitch frequency of vowel /a/. If the waveform is limited around the fundamental frequency, the instantaneous frequency becomes equal to the fundamental frequency. It is not easy, however, to obtain the waveform that contains only the fundamental frequency components of speech, which has a very wide frequency variation.

You can run the programs in Figs. 5.9, 5.10, and 5.11 and try waveforms with different frequency components and other types of waveforms. You will get more findings which are not explained in this section.

5.10 Frequency Modulation

In Sect. 5.9, we defined the instantaneous frequency of a signal and showed how to derive the waveform of the instantaneous frequency. Those examples suggest us that another signal transmission system other than the amplitude modulation is possible, in which signals are transmitted by changing the instantaneous frequency of the carrier. This is exactly the frequency modulation, the main topic of this section.

To introduce the frequency modulation, we will consider a cosine signal, whose frequency is equal to f_c and the phase is made equal to a signal $\Psi(t)$ as shown by the equation below.

$$x(t) = \cos\{2\pi f_c t + \psi(t)\} = \cos[\theta(t)] \tag{5.36}$$

The frequency f_c is the carrier frequency and the instantaneous phase is

$$\theta(t) = 2\pi f_c t + \psi(t) \tag{5.37}$$

The remaining part of the instantaneous phase excluding the term proportional to time ($2\pi f_c t$) is proportional to $\psi(t)$. Therefore, Eq. (5.36) is a cosine wave that has $\psi(t)$ as the phase term. If we can obtain the waveform of the phase term, it becomes possible to transmit and receive the signal. In the amplitude modulation, the amplitude of the carrier is proportional to the modulating signal. In the same manner, since the phase of the carrier is proportional to the modulating signal in Eq. (5.36), it should be referred to as the *phase modulation*. However, it is not easy to obtain $\psi(t)$ from a phase-modulated wave. Contrarily, as we learned in Sect. 5.9, it is much easier to obtain the instantaneous frequency.

Since the instantaneous angular frequency is the time derivative of the instantaneous phase, it is calculated as follows:

$$F(t) = \frac{d\theta(t)}{dt} = 2\pi f_c + \frac{d\psi(t)}{dt} \tag{5.38}$$

The remaining term after subtracting $2\pi f_c$ from the instantaneous angular frequency is proportional to the time derivative of $\psi(t)$. Therefore, if we produce a cosine wave with

$$\psi(t) = \int \varphi(t)dt \tag{5.39}$$

where $\varphi(t)$ is the signal, its instantaneous frequency becomes $2\pi f_c + \varphi(t)$. In this case, since the instantaneous frequency represents the signal, the cosine wave given by Eq. (5.36) can be referred to as the waveform for *frequency modulation.*

Let us consider a case in which $\varphi(t)$ is expressed by the Fourier cosine series.

$$\varphi(t) = \sum_{k=1}^{K} A_k \cos(2\pi f_k t) \tag{5.40}$$

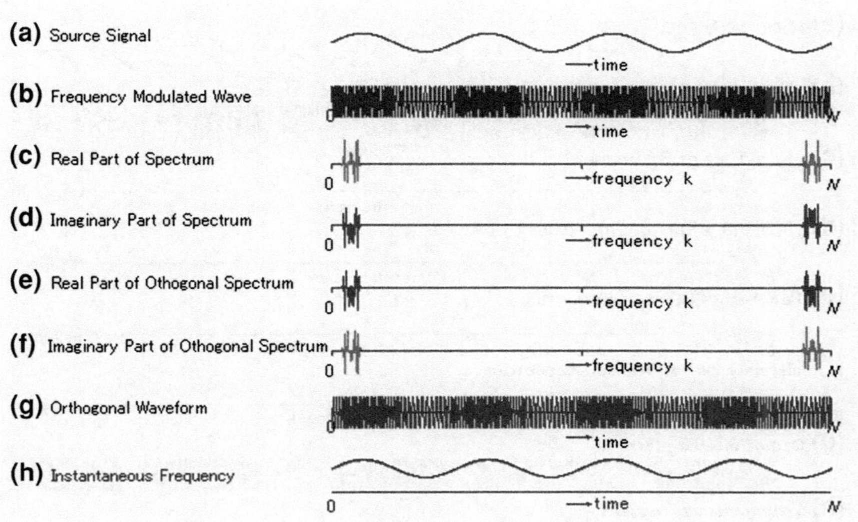

Fig. 5.12 Frequency-modulated wave **b** by sine wave signal (**a**), its spectra (**c**, **d**), orthogonal spectra (**e**, **f**), orthogonal wave (**g**), and demodulated signal (**h**). Animation available in supplementary files under filename E12-12_FM-PM.exe

By integrating this equation, $\psi(t)$ is given as

$$\psi(t) = \sum_{k=1}^{K} \frac{A_k}{2\pi f_k} \sin(2\pi f_k t) \tag{5.41}$$

Then, the frequency modulated wave is expressed as

$$\begin{aligned} x(t) &= \cos\{2\pi f_c t + \psi(t)\} \\ &= \cos[2\pi f_c t + \sum_{k=1}^{K} \frac{A_k}{2\pi f_k} \sin(2\pi f_k t)] \end{aligned} \tag{5.42}$$

Since the instantaneous phase of this wave is $2\pi f_c t + \Psi(t)$, and the instantaneous frequency is $2\pi f_c + \varphi(t)$, if we can obtain the instantaneous frequency in some way, we can recover the signal.

Let us see how $x(t)$ and its spectrum become when $\varphi(t)$ is given by $\varphi(t) = \sin\{2\pi f_S t\}$. The results are shown in Fig. 5.12. Figure 5.12a is the signal $\varphi(t)$ and (b) is the frequency modulated carrier. You can see the frequency variation by the denseness of the waves.

The real and imaginary parts of the spectrum of the frequency modulated carrier are shown by (c) and (d). They show the spreading of the spectrum around the carrier frequency. In the case of the amplitude modulation, there are three line

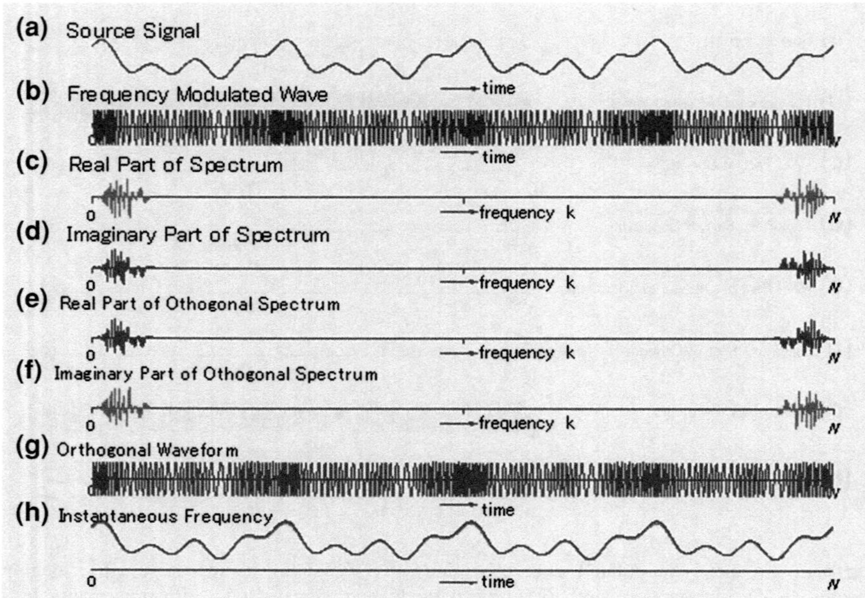

Fig. 5.13 Frequency modulated wave (**b**) by signal (**a**) with three sine components, its spectra (**c**, **d**), orthogonal spectra (**e**, **f**), orthogonal wave (**g**), and demodulated signal (**h**). Animation available in supplementary files under filename E12-13_FM-PM.exe

spectra at f_c (carrier frequency) and at $f_c \pm f_s$ (f_s modulation frequency). However, in the case of frequency modulation, there are many components at $f_c \pm nf_s$, making the spectrum spreading much wider. The spreading of the spectrum is affected much by the amplitude and less by the frequency of the modulating signal. You can check this by running the program attached to Fig. 5.12.

Figure 5.12e, f are obtained by multiplying $-j$ sgn(k) to (c) and (d), respectively. The inverse Fourier transform is (g), the waveform orthogonal to (b). The instantaneous frequency shown by (h) is the angular velocity of the vector which has (b) as its real part and (g) as its imaginary part. Figure 5.12a, h are identical.

Figure 5.13 shows a similar example with a modulating signal made of three sine components. The modulating signal is nicely recovered as expected.

The two examples shown in Figs. 5.12 and 5.13 are the frequency modulations, in which the phase of the carrier is given by the integral of the signal. The demodulated signal (h) matches with the modulating signal (a). In the case of the phase modulation, in which the phase represents the signal, the demodulated signal is a sine wave with the same frequency of the modulating signal if it is made of a single frequency component. If there are more than one, the instantaneous frequency is equal to the derivative of the modulating signal.

If the instantaneous frequency is the derivative of the modulating signal, its integral becomes the modulating signal. This indicates that we can use the phase

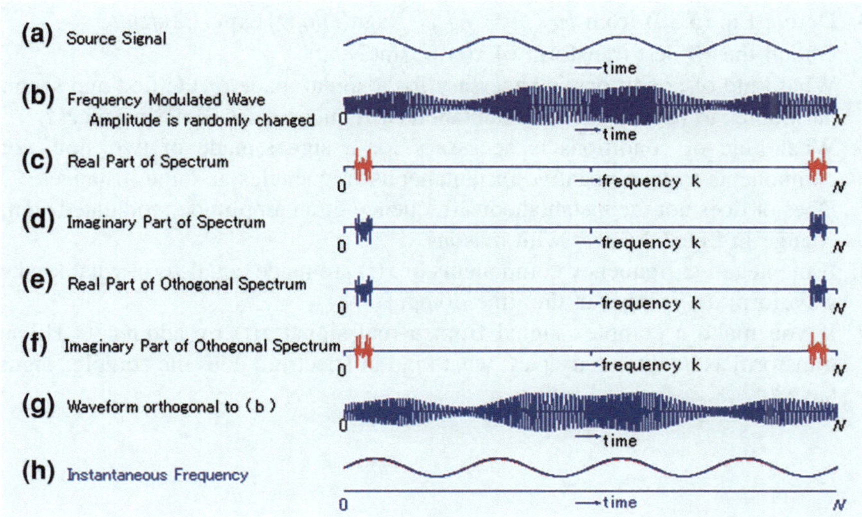

Fig. 5.14 Frequency-modulated carrier with random amplitude change, spectrum, orthogonal wave, and demodulated signal. Animation available in supplementary files under filename E12-14_FM-PM.exe

modulation for communication. That is, after transmitting a phase-modulated carrier, it is demodulated and the resulting instantaneous frequency is integrated at the receiver side. If we call this the *phase modulation*, the difference between the frequency modulation and the phase modulation is only whether the integration of the signal is done before the modulation or after the demodulation. There is an opinion that the latter gives a better result since the integration after the transmission removes the noise contaminated during the transmission.

The frequency modulation based on the instantaneous frequency change of the carrier is not affected by its amplitude change. Figure 5.14 shows it by giving a time variation to the amplitude of the carrier. You can see that the demodulated instantaneous frequency (h) is not affected by the amplitude change of the carrier (b). This is a case for a single frequency signal but the results are the same even for more complex signals.

You can check what were stated in this section by running the programs attached to Figs. 5.12, 5.13, and 5.14.

5.11 Exercise

1. What do you get if you integrate the product of spectrum $X(f)$ and its orthogonal spectrum and $X_{\perp}(f)$ from $-\infty$ to ∞?
2. Derive Eq. (5.26).

3. Derive Eq. (5.30) from $H_O(f) = -j\int_{-\infty}^{+\infty} \mathrm{sgn}(t)h_E(t)\exp(-j2\pi ft)dt$.
4. Obtain the Hilbert transform of $x(t) = \sin(2\pi f_0 t)$.
5. What kind of conditions is necessary for a signal made of the first and second harmonics to take negative instantaneous frequencies at some instances?
6. What kind of conditions is necessary for a signal made of two sinusoidal components to take negative instantaneous frequencies at some instances?
7. Does or does not the instantaneous frequency of an amplitude modulated signal change in time? Answer with reasons.
8. If the negative frequency components of $x(t)$ are made equal zero, what kind of waveform do you get in the time domain?
9. If you make a complex signal from a real signal $x(t)$ by adding its Hilbert transform as its imaginary part, what kind of spectrum does the complex signal have?

Chapter 6
Two-Dimensional Transform

Thus far, we have dealt with one-dimensional signals and related one-dimensional transforms. The reason is that the one-dimensional Fourier transform is easy to understand for learning the basics of Fourier analysis, and in most of application areas, especially in the acoustics area, one-dimensional waves are most common waveforms. However, there is a vast area of application of the Fourier transform in the image processing. Recently, the multimedia technology has become so common and even the acoustics engineers cannot be indifferent to the images and graphics technologies. With these in our mind, we will extend the one-dimensional Fourier transforms to the two-dimensional Fourier transforms.

There is an opinion that we should start from the continuous two-dimensional transforms, but the transforms in the discrete system is easy to understand and the practical applications are for discrete (digitized) systems. Therefore, in this section, we will deal with only the two-dimensional DFT and DCT (*discrete cosine transform*).

6.1 Extension to Two-Dimensional Discrete Fourier Transforms

From here, we deal with two-dimensional waveforms (images), that is, functions with two independent variables. The most common one-dimensional signal is the one with time as its independent variable. For a two-dimensional waveform, both variables cannot be the time. One of them can be a spatial variable, but waveforms with both variables are spatial such as still images or the contours of lands may be easier to conceive. In fact, it can be any function that is defined by two independent variables.

From now on, we will consider two-dimensional DFT by denoting the two-dimensional function by $x(m, n)$, where m and n are integer variables for the horizontal and vertical axes, respectively, of the two-dimensional rectangular

K. Kido, *Digital Fourier Analysis: Advanced Techniques*,
DOI: 10.1007/978-1-4939-1127-1_6,

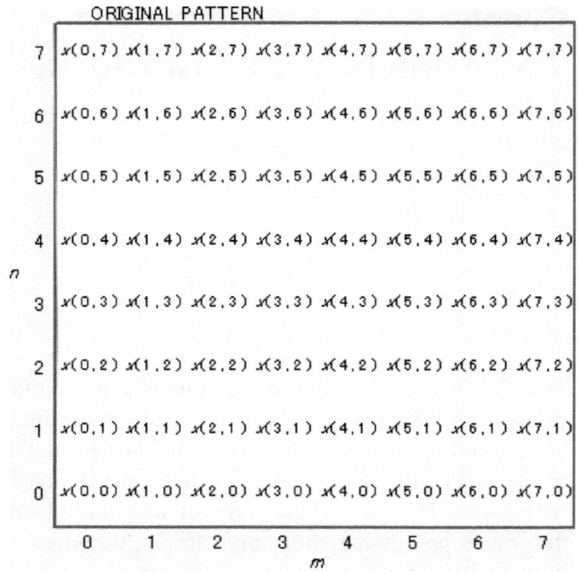

Fig. 6.1 Alignment of $x(m, n)$ with 8×8 independent variables

coordinate system. The ranges of the variables are $0 \le m \le M - 1$ and $0 \le n \le N - 1$.

Figure 6.1 shows columns and rows of values $x(m, n)$ with horizontal and vertical coordinates m and n, respectively, for the case with $M = 8$ and $N = 8$.

In order to derive the two-dimensional DFT of $x(m, n)$, M-point DFT of $x(m, n)$ is performed first with respect to m while keeping n constant. It is expressed as

$$Z(\mu, n) = \sum_{m=0}^{M-1} x(m, n) \exp\left(-j2\pi \frac{\mu \cdot n}{M}\right) \tag{6.1}$$

The new function $Z(\mu, n)$ is shown in Fig. 6.2 with the horizontal axis μ. The order of each row of Fig. 6.2 is the order of the DFT of the sequence of the same row in Fig. 6.1. Even if the $x(m, n)$ is real, values of $Z(\mu, n)$ is generally complex. There is no need for $x(m, n)$ to be real, which is the same as in the one-dimensional case, but since all measured data are real, we will assume that $x(m,n)$ is real.

The next step is the DFT of the sequences $Z(\mu,n)$ in the vertical direction shown in Fig. 6.2. The sequences in the horizontal direction are in the order of discrete frequency and not in the order of spatial coordinate.

The N-point DFT of $Z(\mu, n)$ with respect to n is given by:

$$X(\mu, \nu) = \sum_{n=0}^{N-1} Z(\mu, n) \exp\left(-j2\pi \frac{\nu \cdot n}{N}\right) \tag{6.2}$$

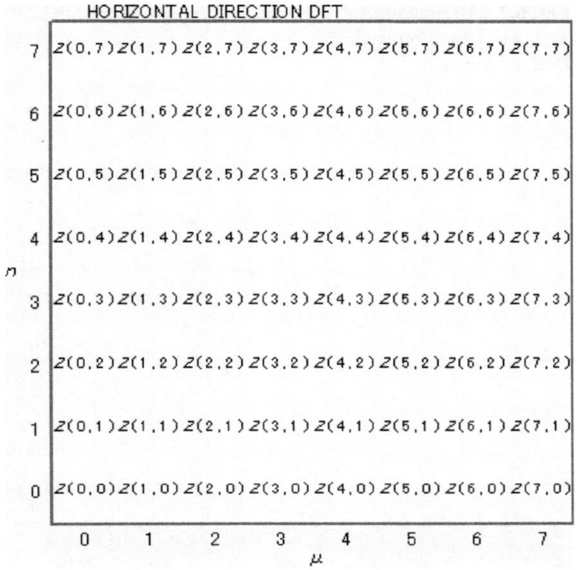

Fig. 6.2 Alignment of $Z(\mu, n)$ with 8×8 independent variables

This is the two-dimensional DFT of $x(m, n)$. Putting Eqs. (6.1) and (6.2) together into one, we have

$$X(\mu, v) = \sum_{n=0}^{N-1} \sum_{m=0}^{M-1} x(m, n) \exp\left(-j2\pi \frac{\mu \cdot m}{M}\right) \exp\left(-j2\pi \frac{v \cdot n}{N}\right) \tag{6.3}$$

As this equation shows, the result is independent on the order of the operations with respect to m and n. The values of $X(\mu, v)$ are shown in Fig. 6.3, in which the coordinates are not the spatial ones but the discrete frequencies. The parameters μ and v are referred to as *spatial frequencies*. The spatial frequency of β means that there are β number periods of sine (or cosine) waves in the spatial coordinate 0 to $M - 1$ or 0 to $N - 1$.

6.2 Two-Dimensional Inverse Discrete Fourier Transform

Following the case of one-dimensional DFT and IDFT, the Two-Dimensional Inverse Discrete Fourier Transform (2D IDFT) of $X(\mu, v)$ is written as:

$$x(m, n) = \frac{1}{MN} \sum_{\mu=0}^{M-1} \sum_{v=0}^{N-1} X(\mu, v) \exp\left(j2\pi \frac{\mu m}{M}\right) \exp\left(j2\pi \frac{vn}{N}\right) \tag{6.4}$$

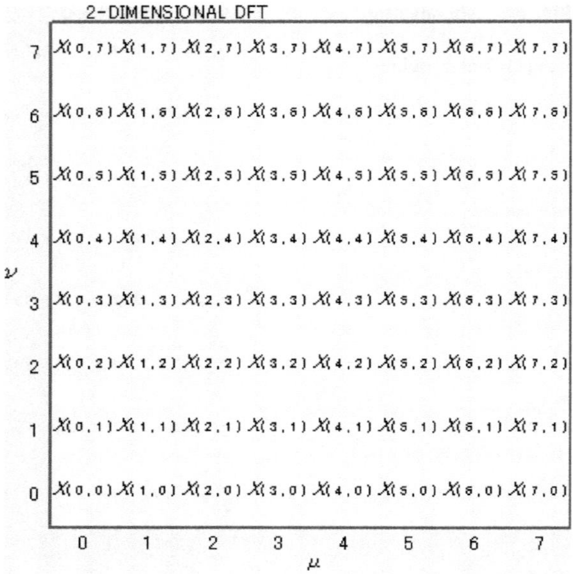

Fig. 6.3 Alignment of $X(\mu, \nu)$ of two-dimensional DFT of $x(m, n)$

Let us check if this gives the inverse transform of Eq. (6.3). It can be done by substituting $X(\mu, \nu)$ given by Eq. (6.3) into Eq. (6.4). However, since it is too lengthy to do at once, we will start from the transform with respect to v. We will use Eq. (6.3), in which m and n are replaced by p and q, respectively.

$$\sum_{v=0}^{N-1} X(v,\mu)\exp\left(j2\pi\frac{vn}{N}\right)$$

$$=\sum_{v=0}^{N-1}\sum_{p=0}^{M-1}\sum_{q=0}^{N-1} x(p,q)\exp\left(-j2\pi\frac{\mu p}{M}\right)\exp\left(-j2\pi\frac{vq}{N}\right)\exp\left(j2\pi\frac{vn}{N}\right)$$

$$=\sum_{p=0}^{M-1}\sum_{q=0}^{N-1} x(p.q)\sum_{\upsilon=0}^{N-1}\exp\left(-j2\pi\frac{v(q-n)}{N}\right)\exp\left(-j2\pi\frac{\mu p}{M}\right)$$

Since the summation of the exponential function with respect to v is zero for $q \neq n$, and N for $q = n$, it is simply given as

$$\sum_{v=0}^{N-1} X(v,\mu)\exp\left(j2\pi\frac{vn}{N}\right) = N\sum_{p=0}^{M-1} x(n,p)\exp\left(-j2\pi\frac{\mu p}{M}\right) \tag{6.5}$$

The calculation with respect to v has been finished. Next is with respect to μ.

$$\frac{N}{MN}\sum_{\mu=0}^{M-1}\sum_{p=0}^{M-1}x(n,p)\exp\left(-j2\pi\frac{\mu p}{M}\right)\exp\left(j2\pi\frac{\mu m}{M}\right)$$

$$=\frac{1}{M}\sum_{p=0}^{M-1}x(n,p)\sum_{\mu=0}^{M-1}\exp\left(-j2\pi\frac{\mu(p-m)}{M}\right)=x(n,m) \quad (6.6)$$

Since the summation of the exponential function with respect to μ is zero for $p \neq m$, and M for $p = m$, the above equation equals $x(m, n)$. It has been proved that Eq. (6.4) is valid as the equation for 2D IDFT.

6.3 Examples of Two-Dimensional DFT and IDFT

Up to now, we have derived equations for two-dimensional DFT and IDFT. In order to make clear what they really mean, numerical examples of 2D DFT ad IDFT of simple 2D images will be shown. All examples are for the cases with $M = N$.

As the easiest example, we will use a perfectly uniform distribution, $x(m, n) = 1$ for both m and n from 0 to 7, as shown by the left top chart of Fig. 6.4. First, the DFT in the horizontal (m) direction is performed. Since $x(m, n)$ is constant in the horizontal direction, the DFT in the m direction has only the DC component. The DC ($\mu = 0$) component is given by $Z(0, n)$ in Eq. (6.1), which is equal to M. The other frequency components are all equal to zero since $\sum_{m=0}^{M}\exp(-j2\pi\mu m/M) = 0, \quad (\mu \neq 0)$. The top middle and right charts of Fig. 6.4 shows that only the real parts of $Z(0, n) = 8$ and all other values are zero.

The next step is the calculation of Eq. (6.2), which is the DFT of $Z(\mu, n)$ in the vertical (n) direction. For $\mu = 0$, since $Z(0, n) = M$ for all n, Eq. (6.2) is equal to MN only when $v = 0$. For $\mu \neq 0$, $Z(\mu, n) = 0$. The real and the imaginary parts of $Z(\mu, v)$ are shown by the bottom middle and right charts of Fig. 6.4, respectively. The real part of $Z(0, 0)$ is 64 and all other real and imaginary values are zero. The sizes of the circles are made larger according to the values, but they are not proportional to the actual sizes (1, 8, and 64).

The IDFT calculated using Eq. (6.4) is shown by the bottom left chart of Fig. 6.4, which is the same with the original image (top left).

The above example with only the DC components is too simple. The Example shown in Fig. 6.5 is the case of 16 × 16 2D DFT/IDFT of a two-dimensional gray image with two and three periods of sine waves in the horizontal and vertical directions, respectively. The original gray image is shown by the top left chart. The filled and hollow circles represent the positive and negative values, respectively, and the radii of the circles represent the magnitudes of the sine waves.

The arrangements of charts in Figs. 6.4 and 6.5 are the same. The original image is shown by the top left chart, and the top middle and right charts show the real and imaginary parts of the DFT with respect to the horizontal axis. Since the

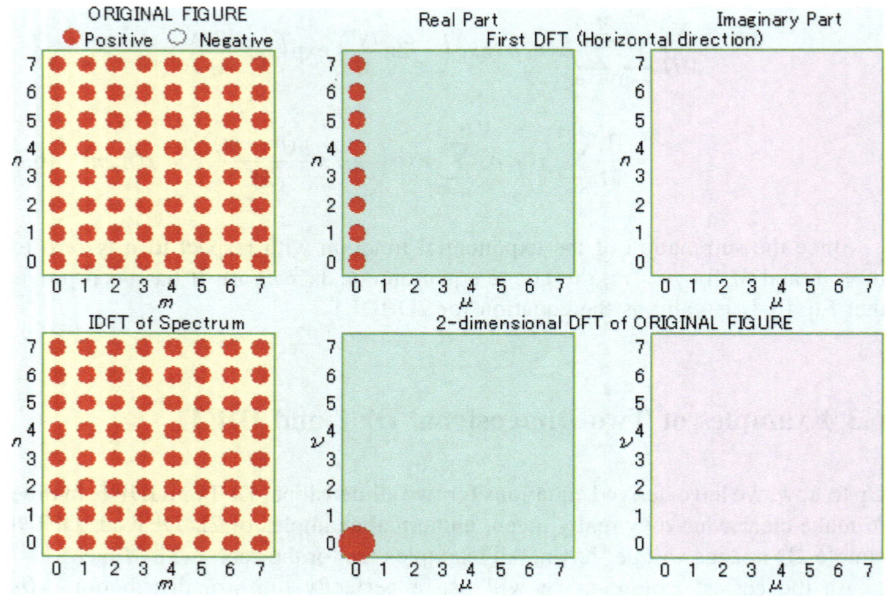

Fig. 6.4 Two-dimensional 8 × 8 DFT/IDFT of an image of a uniform distribution. *Top left* original image, *top middle*, and *right* real and imaginary parts of DFT in the horizontal direction, *bottom middle* and *right* real and imaginary parts of DFT in both directions, *bottom left* IDFT of the DFT. Animation available in supplementary files under filename E13-04_2DDFT.exe

change of the image in the horizontal direction is anti-symmetric and there are two waves, the DFT in the horizontal direction has only the imaginary parts at the discrete frequency 2 and 14. The DFT of $Z(\mu, n)$ with respect to n is shown by the bottom middle (real parts) and right (imaginary parts) charts, which are the 2D DFT of $x(m, n)$. Since there are three imaginary waves in the vertical direction as shown by the top right chart, its DFT has only values at the frequency 3 and 13 in the vertical direction.

In this example, the image is defined by

$$x(n,m) = x_1(m) \cdot x_2(n) \tag{6.7}$$

$$x_1(m) = \sin\left(2\pi \frac{2m}{N}\right)$$

$$x_2(n) = \sin\left(2\pi \frac{3n}{N}\right)$$

Therefore, there should be components at discrete frequencies 2 and 14 on the frequency axis μ, and at discrete frequencies 3 and 13 on the frequency axis v. The

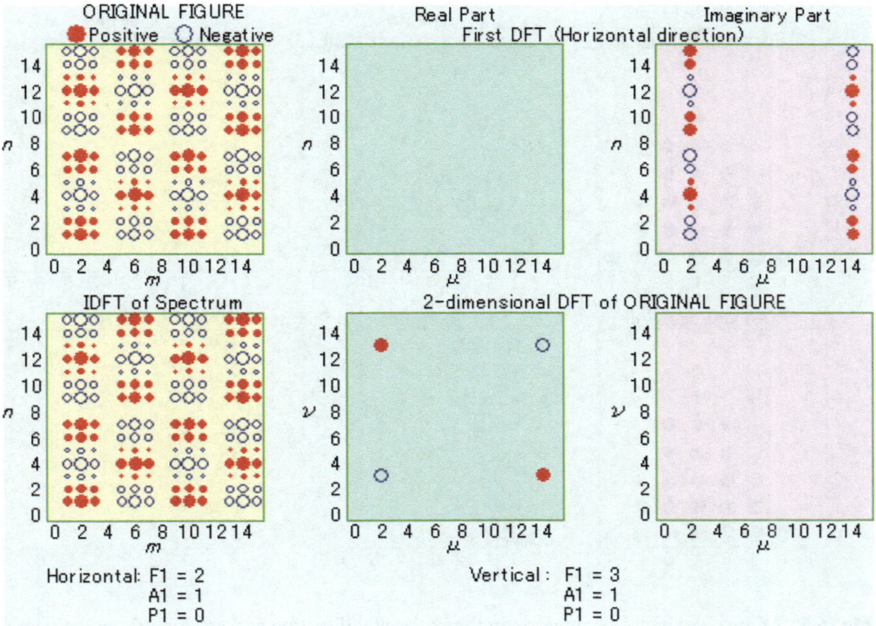

Fig. 6.5 Two-dimensional 16 × 16 DFT/IDFT of an image with two and three waves in the horizontal and vertical axes. Animation available in supplementary files under filename E13-05_2E16_SC.exe

bottom middle chart shows this is exactly the case. Only the real parts have nonzero values in the present case.

The case of Fig. 6.5 is slightly more complex than the case of Fig. 6.4. However, since the dependencies of the image on the two variables are separated, this example is still a special case.

The next example shown in Fig. 6.6 is an image that is symmetric in the horizontal axis and asymmetric in the vertical axis. The arrangements of charts are the same as before. The imaginary components of the DFT in the horizontal axis are completely zero because of the symmetry in the horizontal axis. The DFT in the vertical direction of the real parts show nonzero real and imaginary values as shown by the bottom two right charts, which are the 2D spectra of the original image.

You will notice that the real parts have the symmetry in μ and ν directions, but the imaginary parts have symmetry in the μ direction and antisymmetry in the ν direction. This is because the original image is even in the horizontal direction and its DFT in the same direction has only real values, and DFT of the real values have even real parts and odd imaginary parts. This can be understood from the knowledge you learned in the one-dimensional DFTs.

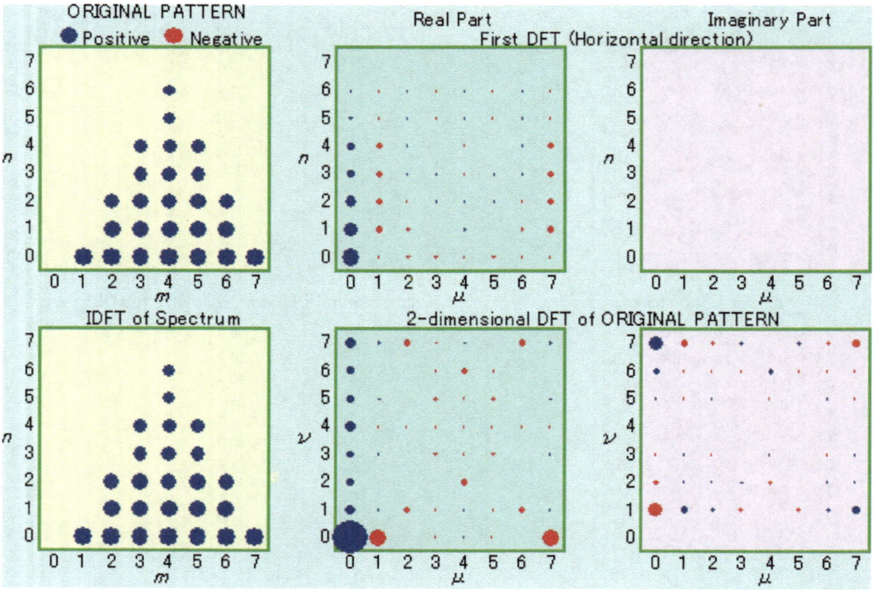

Fig. 6.6 An image which is symmetric in the horizontal direction and asymmetric in the vertical axis and its 2D transform. Animation available in supplementary files under filename E13-06_2D8_TR.exe

6.4 Two-Dimensional Discrete Cosine Transforms (2D DCT)

Let us extend the four kinds of one-dimensional DCTs from DCT-I to DCT-IV to the two-dimensional cases. The same two-dimensional function $x(m, n)$ as in Sect. 6.1, which has M and N points in the horizontal and vertical directions, respectively, is used for the two-dimensional transform.

6.4.1 DCT-1

Let us start from DCT-I. Same as in 2D DFT, the DCT with respect to m is performed while keeping n constant. It can be done by replacing $x(n)$ in Eq. (4.29 of volume I) by $x(m, n)$.

$$Z(\mu, n) = \sqrt{\frac{2}{M}} C_\mu \sum_{m=0}^{M} C_m x(m, n) \cos\left(\pi \frac{\mu m}{M}\right) \tag{6.8}$$

Next, by performing the DCT-I of $Z(\mu, n)$ with respect to n, the 2D DCT $X(v, \mu)$ is obtained.

$$X(\mu, v) = \sqrt{\frac{2}{N}} C_v \sum_{n=0}^{N} C_n Z(\mu, n) \cos\left(\pi \frac{vn}{N}\right) \tag{6.9}$$

The constants C_μ, C_v, C_m, and C_n are given the same way as in Eq. (4.31 of volume I), which is shown here once again.

$$C_k = \begin{cases} 1 & k = 1, 2, \cdots, N-1 \\ 1/\sqrt{2} & k = 0, N \end{cases} \tag{6.10}$$

This coefficient can be commonly used in all equations of the DCT-I.

The 2D DCT-I can be put into one equation if we substitute Eq. (6.8) into $Z(\mu, n)$ in Eq. (6.9), but it will not simplify the situation and will not give us a better insight. Since practical calculation is carried out sequentially from Eq. (6.8) and then (6.9), there is no reason to write down by one equation.

Next is the inverse transform. This can be done by performing the same calculation as Eq. (4.30 of volume I) twice, which are shown below

$$Z(n, \mu) = \sqrt{\frac{2}{N}} C_n \sum_{v=0}^{N} C_v X(v, \mu) \cos\left(\pi \frac{nv}{N}\right) \tag{6.11}$$

and

$$x(n, m) = \sqrt{\frac{2}{M}} C_m \sum_{\mu=0}^{M} C_\mu Z(n, \mu) \cos\left(\pi \frac{m\mu}{M}\right) \tag{6.12}$$

.

The formulae for the DCT-I and IDCT-I have been given. One example of analysis by the DCT-I is shown by Fig. 6.7. The image is in this case is uniform everywhere. For this image, the 2D DFT gives only the DC component, which is located at the left bottom corner. The DCT-I, however, have small values almost everywhere as well as the large DC component.

6.4.2 DCT-II

In this case, the forward transform is done by performing Eq. (4.24 of volume I) twice. The transform of $x(m, n)$ with respect to m is performed as follows:

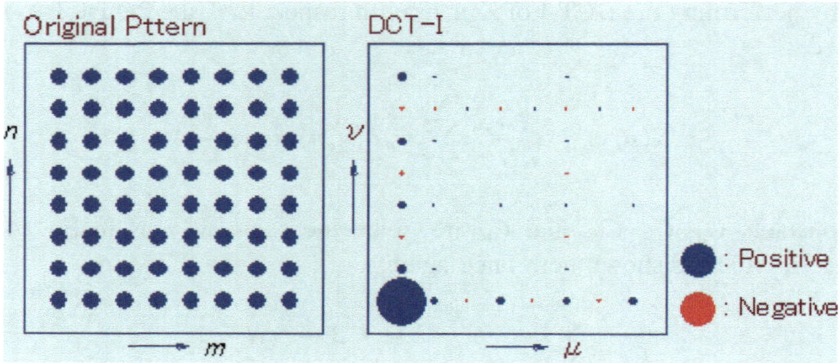

Fig. 6.7 Result of application of the DCT-I to a uniform image. Animation available in supplementary files under filename E13-07_2dDCT-I.exe

$$Z(\mu,n) = \sqrt{\frac{2}{M}} C_\mu \sum_{m=0}^{M-1} x(m,n) \cos\left(\pi \frac{\mu(m+0.5)}{M}\right) \tag{6.13}$$

where

$$C_0^2 = 1/2,\ C_k = 1\ :\ k > 0. \tag{6.14}$$

These coefficients are also common for all equations in DCT-II and IDCT-II. By performing the DCT of $Z(\mu, n)$ with respect to n, 2D DCT $X(\mu, \nu)$ is obtained.

$$X(\mu,\nu) = \sqrt{\frac{2}{N}} C_\mu \sum_{n=0}^{N-1} Z(\mu,n) \cos\left(\pi \frac{\nu(n+0.5)}{N}\right) \tag{6.15}$$

Next, the process of the inverse transform operation is given. In this case, the operation of Eq. (4.25 of volume I) n is performed twice. At first, $Z(\mu, n)$ is obtained from $X(\mu, \nu)$ by the transform regarding ν.

$$Z(\mu,n) = \sqrt{\frac{2}{N}} \sum_{\nu=0}^{N-1} C_\nu X(\mu,\nu) \cos\left(\pi \frac{\nu(n+0.5)}{N}\right) \tag{6.16}$$

The same operation is carried out on $Z(\mu, n)$ to get $x(m, n)$

$$x(m,n) = \sqrt{\frac{2}{M}} \sum_{\mu=0}^{M-1} C_\mu Z(\mu,\nu) \cos\left(\pi \frac{\mu(m+0.5)}{M}\right) \tag{6.17}$$

The result of the 2D DCT-II application to the same image as in the last example is shown in Fig. 6.8. Contrary to the DCT-I, values other than DC

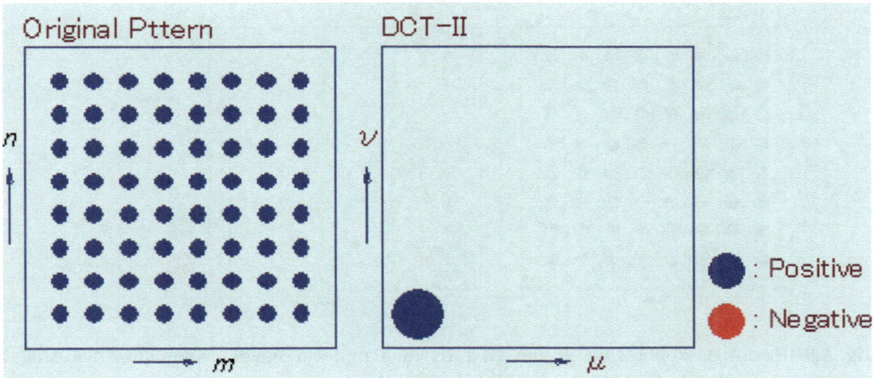

Fig. 6.8 Result of application of the DFT-II to a uniform image. Animation available in supplementary files under filename E13-08_2dDCT-II.exe

component are completely zero, which is the same as the case of DFT. This can be easily understood from the result of Fig. 4.17.

6.4.3 DCT-III

Since the procedure is the same as before, only the results will be shown. Forward transform:

$$Z(\mu, n) = \sqrt{\frac{2}{M}} \sum_{m=0}^{M-1} C_m x(m, n) \cos\left(\pi \frac{(\mu + 0.5)m}{M}\right) \tag{6.18}$$

$$X(\mu, v) = \sqrt{\frac{2}{N}} \sum_{n=0}^{N-1} C_n Z(\mu, n) \cos\left(\pi \frac{(v + 0.5)n}{N}\right) \tag{6.19}$$

where

$$C_0^2 = 1/2,\ C_k = 1\ :\ k > 0 \tag{6.20}$$

Inverse transform:

$$Z(\mu, n) = \sqrt{\frac{2}{N}} C_n \sum_{v=0}^{N-1} X(\mu, v) \cos\left(\pi \frac{(v + 0.5)n}{N}\right) \tag{6.21}$$

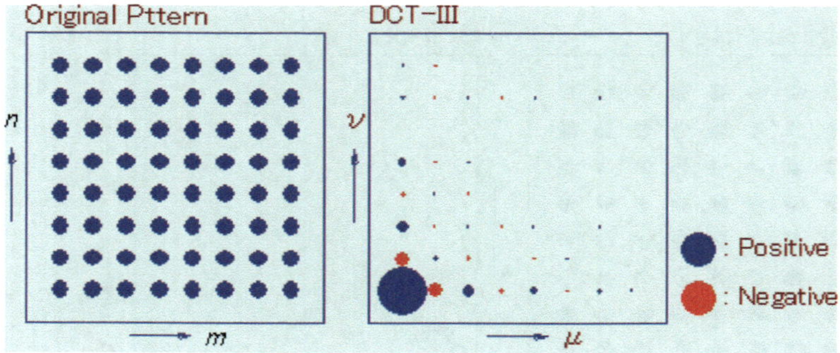

Fig. 6.9 Result of application of the DFT-III to a uniform image. Animation available in supplementary files under filename E13-09_2dDCT-III.exe

$$x(m,n) = \sqrt{\frac{2}{M}} C_m \sum_{\mu=0}^{M-1} Z(\mu,n) \cos\left(\pi \frac{(\mu+0.5)m}{M}\right) \tag{6.22}$$

Fig. 6.9 shows the result of the application of DCT-III to the uniform image. As you can imagine from Fig. 4.17, the values other than the DC component take larger values than the previous example.

6.4.4 DCT-IV

Forward transform:

$$Z(\mu,n) = \sqrt{\frac{2}{M}} \sum_{m=0}^{M-1} x(m,n) \cos\left\{\pi \frac{(m+0.5)(\mu+0.5)}{M}\right\} \tag{6.23}$$

$$X(\mu,\nu) = \sqrt{\frac{2}{N}} \sum_{n=0}^{N-1} Z(\mu,n) \cos\left\{\pi \frac{(n+0.5)(\nu+0.5)}{N}\right\} \tag{6.24}$$

Inverse transform:

$$Z(\mu,n) = \sqrt{\frac{2}{N}} \sum_{\nu=0}^{N-1} X(\mu,\nu) \cos\left\{\pi \frac{(\nu+0.5)(n+0.5)}{N}\right\} \tag{6.25}$$

$$x(m,n) = \sqrt{\frac{2}{M}} \sum_{\mu=0}^{M-1} Z(\mu,n) \cos\left\{\pi \frac{(\mu+0.5)(m+0.5)}{M}\right\} \tag{6.26}$$

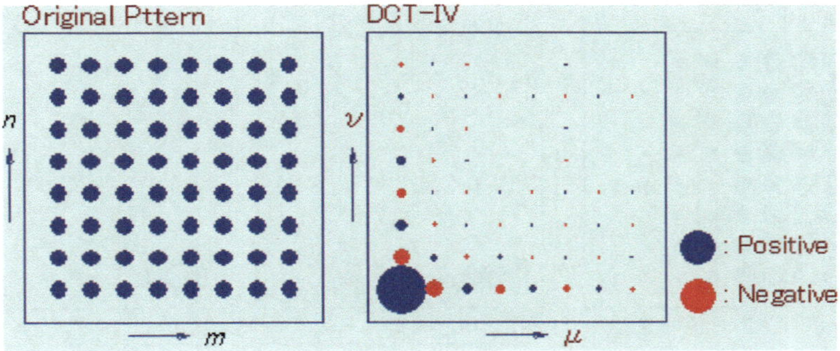

Fig. 6.10 Result of application of the DFT-IV to a uniform image. Animation available in supplementary files under filename E13-10_2dDCT-IV.exe

The equations shown above indicate that results are independent of the order of the transform with respect to m or n and μ or ν. With regard to the other figures, you can run the programs and find features of the four types of transforms. Those features will be explained in the following two sections.

Figure 6.10 shows the result of the application of DCT-III to the uniform image. In this case, the spreading of the spectrum is large.

6.5 DCT of Gray Images

Thus far, the DCT results of the uniform image have been shown. Figures 6.11, 6.12, 6.13, and 6.14 show results of application to other gray images. Four 8×8 images are shown by the left top charts in the figures. The four transforms of the figures are shown by the other four charts in the figures. The 8×8 gray image patterns are shown by the sizes of the circles which are proportional to the degree of the density of the gray color. The absolute values of the spectra are also indicated proportionally to the sizes of the circles. The filled (blue) and hollow (red) circles are for positive and negative spectra, respectively.

Figures 6.11, and 6.12 are for the images, in which the density decreases linearly in the horizontal direction and in the diagonal direction toward the low right corner, respectively. Even the images have spatial variations, their DCT spectra are concentrated near the low frequency range. Especially, those of DCT-II are remarkable. This is because the image has a very smooth variation and contains small high frequency components. If the image contains high frequency variations as shown in Fig. 6.13, those are reflected in the DCT spectra.

In the above three examples, the images have smooth density changes. Let's check another image that has an abrupt density change along the horizontal axis. You can guess that this kind of image has large high frequency components from the examples of the one-dimensional DFTs discussed before. Figure 6.14 shows an

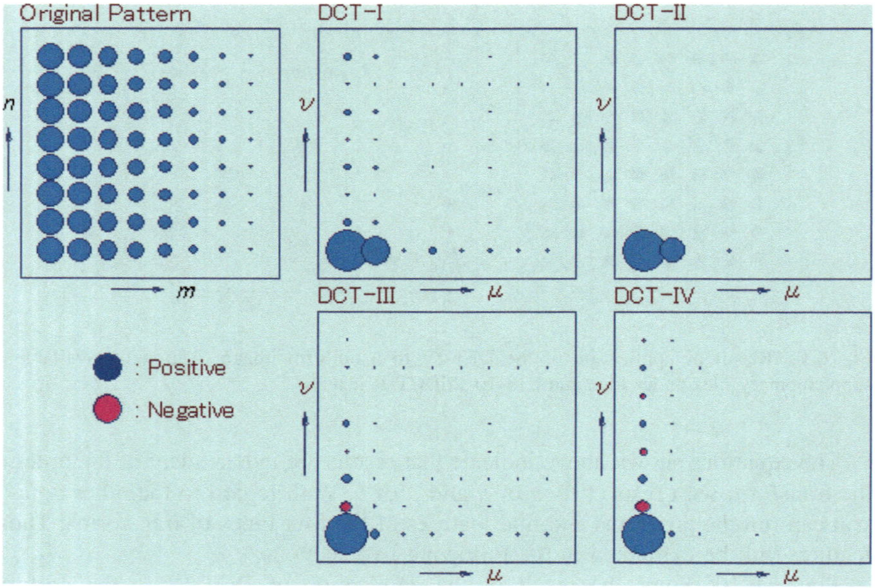

Fig. 6.11 2D DCTs of the 8 × 8 gray image whose density decreases linearly in the horizontal direction. Animation available in supplementary files under filename E13-11_2dDCT.exe

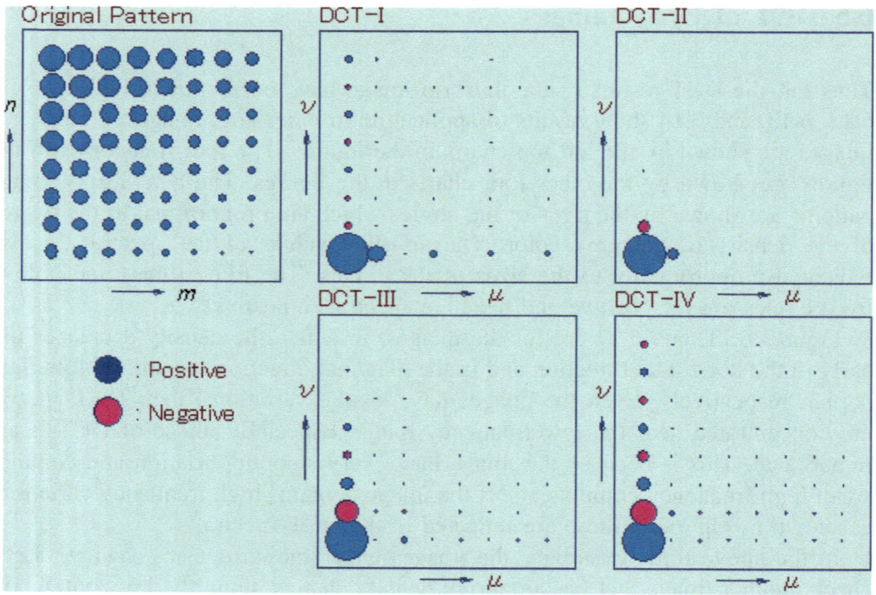

Fig. 6.12 2D DCTs of the 8 × 8 gray image whose density decreases linearly in the diagonal direction. Animation available in supplementary files under filename E13-12_2dDCT.exe

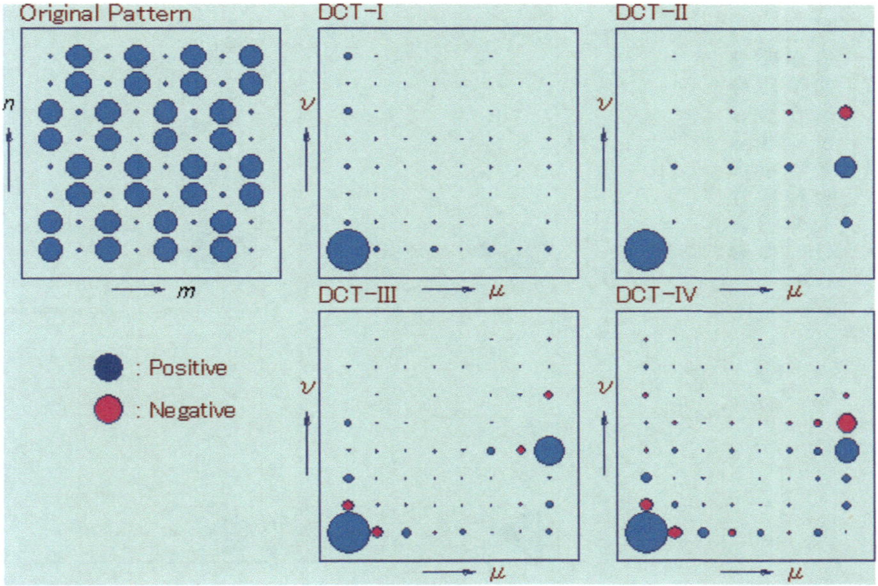

Fig. 6.13 2D DCTs of the 8 × 8 gray image whose density changes with four and two periods in the horizontal and vertical directions, respectively. Animation available in supplementary files under filename E13-13_2dDCT.exe

example of DCTs of the image that has density 1 in the left half and 0 in the right half. In this example, DCTs take the largest values at $\mu = \nu = 0$ because the image pattern takes only positive and 0 values. If the image takes +1 and −1 in the left and right half, respectively, the value of the DCT-II at $\mu = \nu = 0$ becomes 0, but not necessarily so in the case of the other three DCTs.

Because of the space limitation, examples of other interesting images cannot be listed in this here. You can run the program attached to the figures and check the results of DCTs for other images.

6.6 Data Compression

It is not surprising that a reconstructed image may have a small deterioration if only low level frequency components of the DCTs are neglected. Let us discuss the commonly used data compression technique that utilizes this property.

The examples from Figs. 6.7, 6.8, 6.9, 6.10, 6.11, 6.12, and 6.14 (except for Fig. 6.13) show that the values of DCTs at points far from the point $\mu = \nu = 0$ are very small. This suggests that if only those values close the point $\mu = \nu = 0$ are kept, the reconstructed images may be close to the originals.

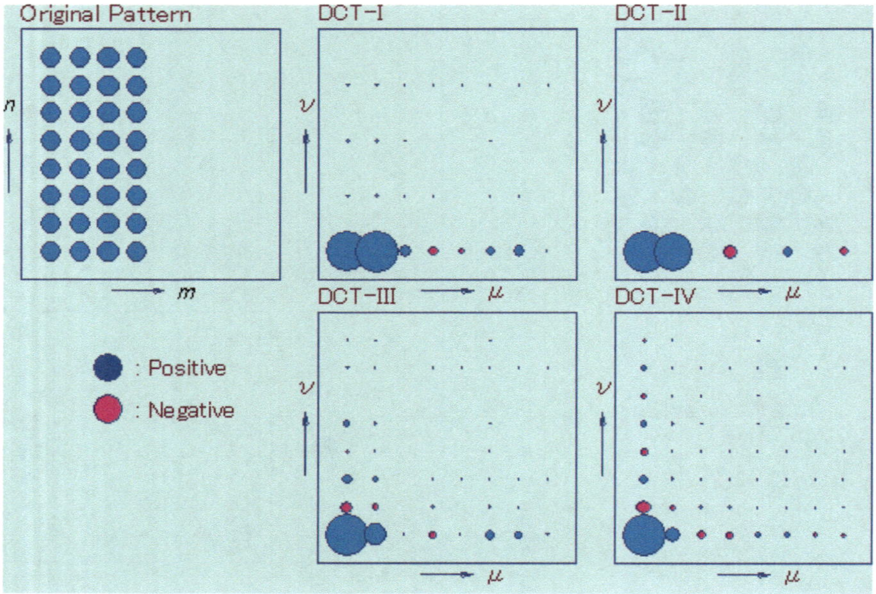

Fig. 6.14 2D DCTs of the 8 × 8 gray image whose density changes abruptly only in the horizontal direction. Animation available in supplementary files under filename E13-14_2dDCT.exe

However, the image made of 8 × 8 points are too small. The data compression is really necessary when a complex image with a large amount of data must be sent or recorded and reproduced. Therefore, we will consider a case when a large image is divided into many small subimages and those are sent to a receiver. At the receiver side, the original image is reconstructed using those subimages. If the subimage is made of 8 × 8 point data, the image variation in it may be mostly very small.

If the variation is small, the spectral distribution of each subimage should be concentrated in the low frequency region near the DC component. The gray color information made of a 64-point subimage is transformed by one of the four DCTs and only the low frequency components close to the left bottom corner are sent. The receiver can reconstruct each subimage using the received data and by letting the unreceived data be zero. The receiver then can reconstruct the whole image by assembling all subimages in predetermined order. By reconstructing the original image by this way, the DCTs can be useful tools for the data compression for image communications and storages.

The region, data of which should be reserved, is near the left bottom corner of the 2D spectrum. Let us discuss how large this region should be. Since the DCT-II performs best in the sense that it has the largest spectrum concentration among the four DCTs, we will consider only the performance of the DCT-II.

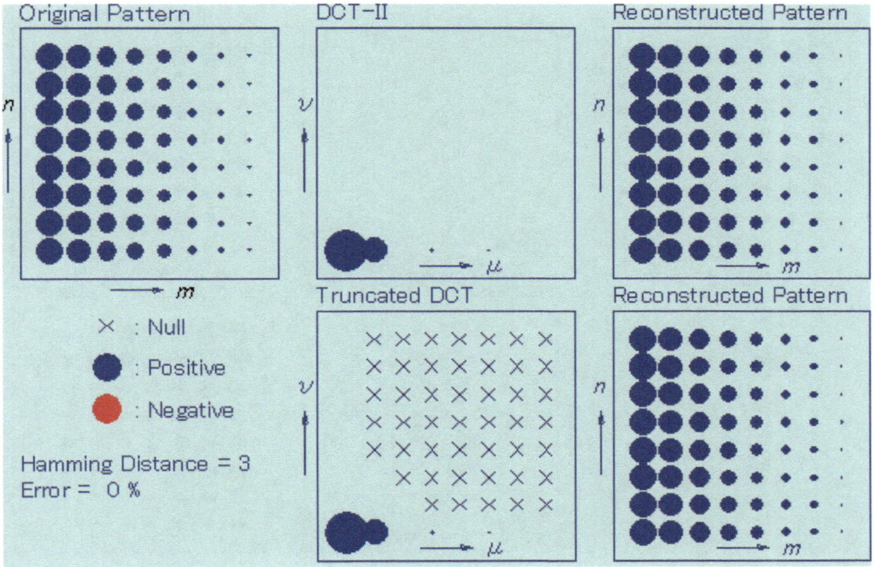

Fig. 6.15 2D DCT-II of the 8 × 8 gray image whose density decreases linearly in the horizontal direction and its IDFT-IIs. The *top middle chart* is the DFT-II of the original image. The symbol × in the *bottom middle chart* shows the spectra which are made equal to zero. The *bottom right chart* shows its IDCT. Animation available in supplementary files under filename E13-15_2dDCT.exe

The spectral distributions of DCT-II in Figs. 6.7, 6.8, 6.9, 6.10, 6.11, 6.12, 6.13, and 6.14 tell us that the large spectral values are concentrated within the range in which μ or v is close to zero or $\mu + v$ is small. For this reason, we will discuss the smallest limit of the Hamming distance, within which the data is reserved and otherwise discarded, where the *Hamming distance* from the origin (0, 0) to the point (μ, v) is defined here as $\mu + v$ instead of the Euclidean distance $\sqrt{\mu^2 + v^2}$. Let us consider first the image shown by the top left chart in Fig. 6.15 (same as the one in Fig. 6.11). The density of this image decreases linearly from the maximum to zero in the horizontal direction. The 2D spectrum is shown by the top middle chart and its IDCT by the top right chart. The last one is of course the same with the original image. The spectrum $X(v, \mu)$ is zero for $v \geq 1$ and $X(\mu, 0)$ is very small for $\mu > 3$. Therefore, even if the values that satisfy the condition $\mu + v > 3$ (shown by X in the bottom left chart) are made equal to zero, the IDCT shown by the bottom right chart is very close to the original image. The error given by the ratio of the squared sum of the difference of corresponding component values to the squared sum of the original image components is 0.03 %. The image close to the original can be reconstructed only by 10-point data instead of sending 64-point data.

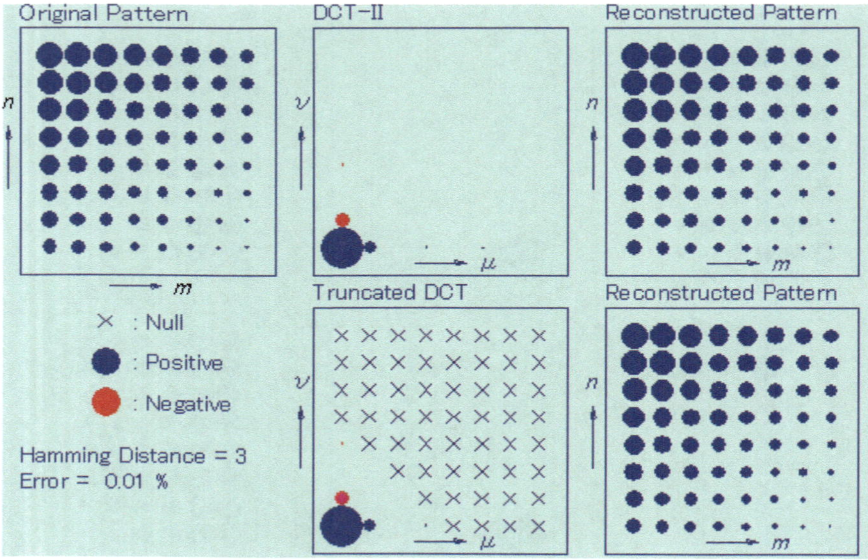

Fig. 6.16 Image with variation in the diagonal direction and results of the same operation as in Fig. 1.15. Animation available in supplementary files under filename E13-16_2dDCT.exe

The image in Fig. 6.15 is a special case which has variation only in the horizontal direction. The next example shown by Fig. 6.16 (same as in Fig. 6.12) is the image which has variation in the diagonal direction. In this example, the error is also very small. From these it is expected that a large data reduction can be achieved.

The next example is the image which has the largest high frequency components among images that have been discussed. The image has the maximum density 1 for $x < 4$ and zero for $x \geq 4$. The image and the results of the same operation as before are shown in Fig. 6.7. Since the spectrum components $X(v, \mu) = 0$ for $v \geq 1$, and $X(0, \mu)$ for $\mu \geq 4$ are very small, it is expected that the reconstructed image may be similar to the original one. The reconstructed image is shown by the bottom right chart which has an error of 3.88 %. Whether this error is allowed or not depends on the purpose of application.

If the limit of the Hamming distance $(\mu + v)$ that determines the null data area shown by $\times$ in Figs. 6.15, 6.16, and 6.17 is made larger, the reconstructed image gets closer to the original image because the spectral difference between the two gets smaller.

As stated above, $X(v, \mu)$ of the image in Fig. 6.17 is equal to 0 for $v \geq 0$. If the data $X(0, \mu)$ (the bottom row) and $X(v, 0)$ (= 0, the left column) are used for the reconstruction, it is obvious that the identical image is obtained as shown in Fig. 6.18.

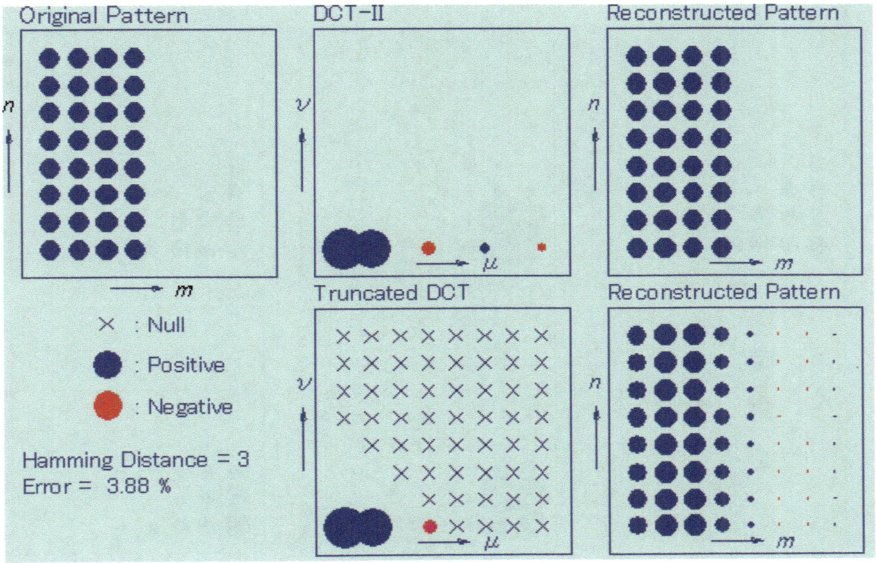

Fig. 6.17 Image with stepwise variation in the horizontal direction and results of the same operation as in Figs. 13.15 and 13.16. Animation available in supplementary files under filename E13-17_2dDCT.exe

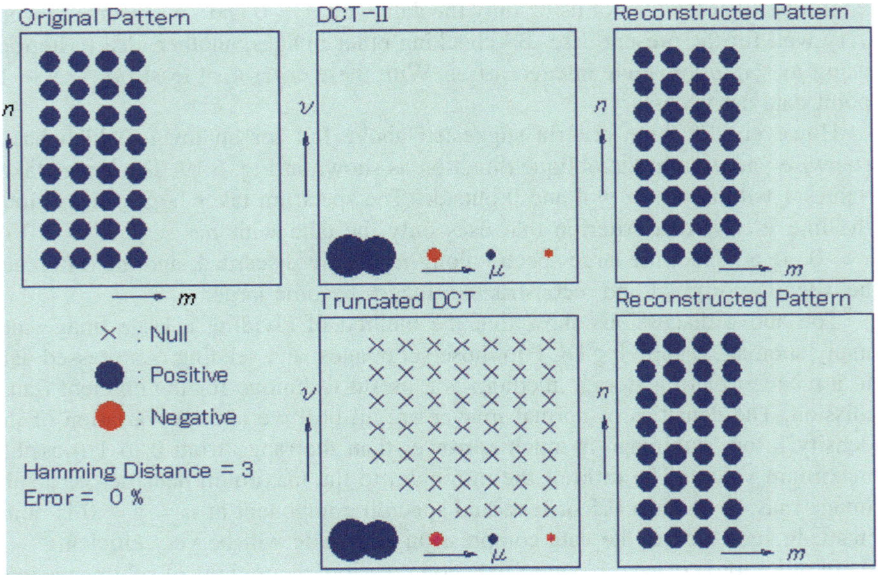

Fig. 6.18 Image which has a stepwise variation in the horizontal direction and its reconstructed image using only the data of the *bottom row* and the *left column*. Animation available in supplementary files under filename E13-18_2dDCT.exe

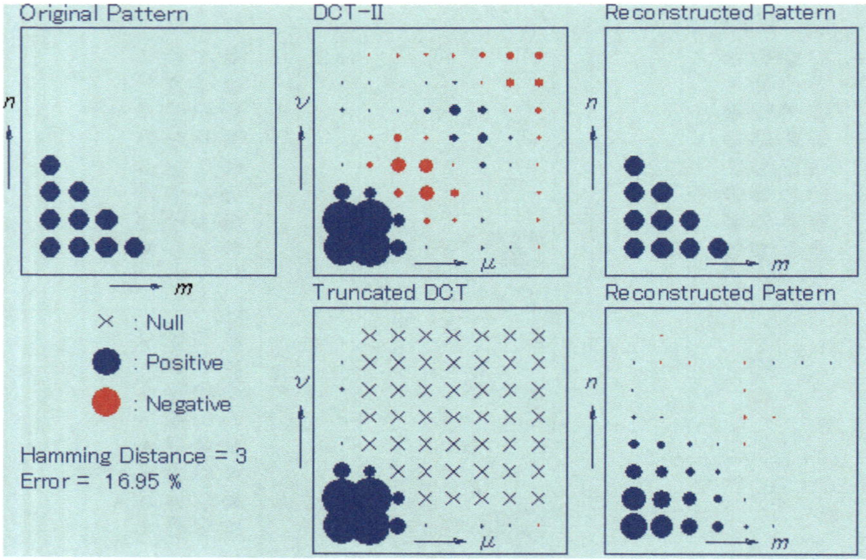

Fig. 6.19 Image which has a stepwise variation in the diagonal direction and its reconstructed image by the same operation as in Fig. 6.15. Animation available in supplementary files under filename E13-19_2dDCT.exe

It looks that the idea of using only the data with $\mu = 0$ and $\nu = 0$ seems to work very well for the present case. By checking other images, another idea (criterion) using $\mu\nu \leq q$, q : positive integer, arises. With these criteria, at least $(M + N - 1)$ point data are necessary.

However, the three criteria suggested above fail for an image which has a stepwise variation in the oblique direction as shown in Fig. 6.19. The image takes values 1 within $m + n \leq 4$ and 0 outside. The spectrum takes large values along the line $\mu = \nu$. If a criterion that uses only the data with $\mu + \nu \leq 3$, $\mu = 0$ or $\nu = 0$ are adopted, the large spectra along $\mu = \nu$ are discarded, and the difference between the original and reconstructed images become large.

The above discussions show that the method of dividing a large image into many submages, applying DCT to those subimages, and sending compressed data to a receiver or to a storage medium is a useful technique for the efficient transmission. The densities of normal images are all positive and the variation of the density in one subimage is much narrower than the range from 0 to 1 (possible maximum value). The ratio of the minimum to the maximum densities of a subimage may be at most 0.5 or less, and spectral component at $\mu = \nu = 0$ is dominant. In such a case, the data compression technique will be very efficient.

In order to increase the rate of data compression, the regions of subimages may change one-by-one. In this case, information on the discarded data region of each subimage must be sent, but if the total amount of information is reduced, it is worth to do so.

The programs attached to the figures allow you to check differences of DCT-I to DCT-IV, to try different images, and to change discarded data regions.

6.7 Exercise

1. Obtain a $N \times N$ two-dimensional DFT of a function x(m, n), where $x(0, 0) = 1$ and $x(m, n) = 0$ otherwise.
2. Obtain a $N \times N$ two-dimensional DFT of a function x(m, n), where $x(m, 0) = 1$ for $m = 0, 1, \ldots N - 1$ and $x(m, n) = 0$ for $n = 1, 2, \ldots, N - 1$.
3. What do you get in the previous problem, if the ranges of m and n are 0 to $N - 1$, and 0 to $M - 1$, respectively?
4. Obtain a $N \times N$ two-dimensional DFT of a function $x(m, n)$, where $x(0, 0) = 2$ and $x(m, n) = 1$ otherwise.
5. Derive the two-dimensional DFT by performing the DFT in the vertical direction first and then the DFT in the horizontal direction.
6. Derive the $N \times N$ DFT and IDFT.
7. What is the answer if the DFT is replaced by DCT in question 1?
8. What is the answer if the DFT is replaced by DCT in question 2?
9. In what situation is the data compression possible in gray image transmissions?
10. Why are there no imaginary components in the frequency domain functions of two-dimensional DCTs such as in Fig. 6.8?

Appendix

Appendix 1A Rewriting the Convolution Formula

Ken'iti Kido

$$y(n) = \sum_{p=0}^{m-1} x(n-p)h(p) \qquad (1.1) \qquad\qquad (1A.1)$$

The parameter p is transformed as

$$r = n - p \qquad (p = n - r)$$

Then, for $p = m - 1$, $r = n - m + 1$, and for $p = 0$, $r = n$. Since $n - m + 1 < n$, by reversing the order of summation, Eq. (1A.1) is rewritten as

$$y(n) = \sum_{r=n-m+1}^{n} x(n-r)h(r) \qquad (1.2) \qquad\qquad (1A.2)$$

Appendix 1B Single Degree-of-Freedom Transfer System

The mechanical system shown in Fig. 1B.1a and the electrical system shown in Fig. 1B.1b has the same impulse response used in Fig. 1.5a in Sect. 1.4. In both systems, the input is x and the output is v. One example of practical systems is the mechanical system represents the relation between the car body displacement v and the road surface displacement (undulation of the road) x.

For both mechanical and electrical systems, the relations between x and v are described by the following two differential equations, which have the identical forms.

K. Kido, *Digital Fourier Analysis: Advanced Techniques*,
DOI: 10.1007/978-1-4939-1127-1,

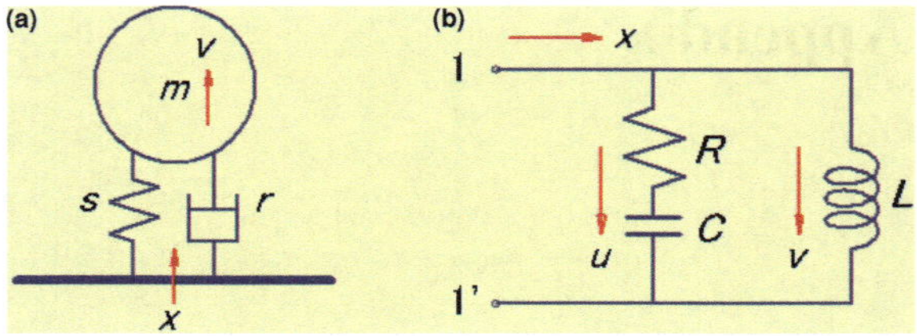

Fig. 1B.1 Single degree-of-freedom transfer systems that have the impulse response used in Fig. 1.5 (**a**). **a** mechanical system, **b** electrical system

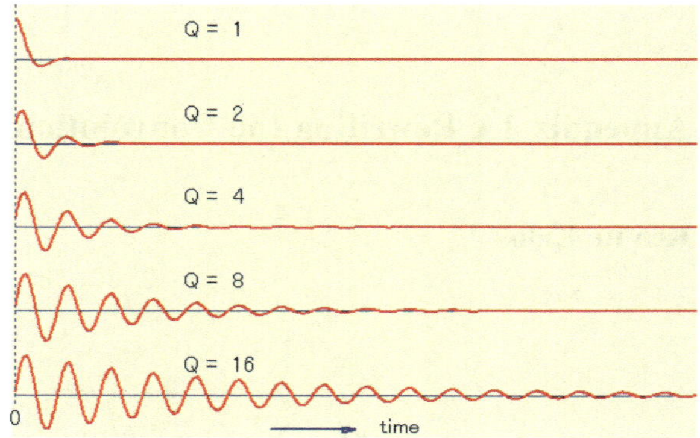

Fig. 1B.2 Impulse responses for various Qs of the transfer systems shown in Fig. 1B.1

$$r\frac{dx}{dt} + sx = m\frac{d^2v}{dt^2} + r\frac{dv}{dt} + sv \qquad \text{mechanical system} \tag{1B.1}$$

$$R\frac{dx}{dt} + \frac{x}{C} = L\frac{d^2v}{dt^2} + R\frac{dv}{dt} + \frac{v}{C} \qquad \text{electrical system} \tag{1B.2}$$

Solving these equations under the condition $2\sqrt{ms} > r$ or $2\sqrt{L/C} > R$, the impulse response is given by

$$h(t) = 2\alpha\, e^{-\alpha t}\left\{\cos\beta t + \frac{2Q^2 - 1}{\sqrt{4Q^2 - 1}}\sin\beta t\right\} \tag{1B.3}$$

where

$$\omega_0 = 2\pi f_0 = \sqrt{\frac{s}{m}} = \frac{1}{\sqrt{LC}}: \quad \text{resonance frequency of forced vibration} \tag{1B.4}$$

$$Q = \frac{\omega_0 m}{r} = \frac{\omega_0 L}{R} > \frac{1}{2}: \qquad \text{quality factor} \tag{1B.5}$$

$$\alpha = \frac{\omega_0}{2Q} = \frac{r}{2m} = \frac{R}{2L}: \qquad \text{damping coefficient} \tag{1B.6}$$

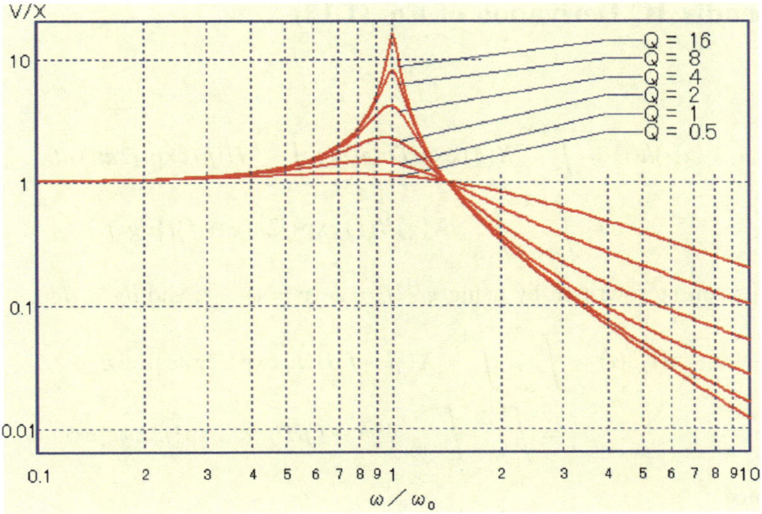

Fig. 1B.3 Frequency response functions of the transfer systems shown in Fig. 1B.1

$$\beta = \omega_0\sqrt{1 - \frac{1}{4Q^2}}: \qquad \text{resonance frequency of free vibration} \tag{1B.7}$$

Figure 1B.2 shows impulse responses for various Qs. The attenuation is inversely proportional to Q.

The transfer function (Fourier transform of the impulse response) of the system is given by

$$\frac{V}{X} = \frac{1 - j\frac{\omega_0}{\omega}Q}{1 + jQ\left(\frac{\omega}{\omega_0} - \frac{\omega_0}{\omega}\right)} = \frac{1 - j\frac{f_0}{f}Q}{1 + jQ\left(\frac{f}{f_0} - \frac{f_0}{f}\right)} \tag{1B.8}$$

where X and V are the complex amplitudes of x and v, i.e., $x = x(t) = X\exp(j2\pi ft)$ and $v = v(t) = V\exp(j2\pi ft)$.

The frequency response of the transfer system is shown in Fig. 1B.3. Let's consider the analogy of the transfer system to the car body vibration. In the low frequency region below the resonance frequency (ω/ω_0), the car body moves the same way as the road surface. At the resonance frequency, the body vibrates with the amplitude of approximately Q times of the road surface undulation amplitude. The body vibration decreases in proportion to the frequency above the resonance frequency. As a vibration isolation system, the resonance frequency must be made much lower than the low end of the frequency range of isolation.

Appendix 1C Derivation of Eq. (1.12)

$$
\begin{aligned}
x(t)h(t) &= \int_{-\infty}^{+\infty} X(g)\exp(j2\pi gt)dg \int_{-\infty}^{+\infty} H(f)\exp(j2\pi ft)\,df \\
&= \int_{-\infty}^{+\infty}\int_{-\infty}^{+\infty} X(g)H(f)\exp\{j2\pi(g+f)t\}dgdf
\end{aligned}
$$

This can be rewritten by using $g+f=\phi$ ($g=\phi-f$, and $dg=d\phi$),

$$
\begin{aligned}
x(t)h(t) &= \int_{-\infty}^{+\infty}\int_{-\infty}^{+\infty} X(\phi-f)H(f)\exp\{j2\pi\phi t\}d\phi df \\
&= \int_{-\infty}^{+\infty}\int_{-\infty}^{+\infty} X(\phi-f)H(f)df\exp\{j2\pi\phi t\}d\phi
\end{aligned}
$$

Since

$$
\int_{-\infty}^{+\infty} X(\phi-f)H(f)df = X(\phi)*H(\phi)
$$

we have

$$
x(t)h(t) = \int_{-\infty}^{+\infty} [X(\phi)*H(\phi)]\exp(j2\pi\phi t)d\phi = \mathrm{IFT}[X(\phi)*H(\phi)] \qquad (1.12) \quad (1C.12)
$$

Appendix 2A Cross-Correlations Between Input and Output of Signals Contaminated with Noises

When an observed input signal $x(n)$ is contaminated with noise $n_1(n)$, and the output $y(n)$, which is given by the convolution of the input and the impulse response, is contaminated by noise $n_2(n)$, Eq. (2.7) is modified as:

$$
y(n) = \sum_{p=0}^{K-1} x(n-p)h(p) + n_2(n) \tag{2A.1}
$$

Then Eq. (2.5) is rewritten by replacing the input by $x(n) + n_1(n)$ as

$$
r_{xy}(m) = \frac{1}{L\sigma_x\sigma_y}\sum_{n=0}^{L-1}\{x(n)+n_1(n)\}\left\{\sum_{p=0}^{K-1}x(n+m-p)h(p)+n_2(n+m)\right\}
$$

This is further rewritten as

$$r_{xy}(m) = \frac{1}{L\sigma_x\sigma_y}\sum_{p=0}^{K-1} h(p)\sum_{n=0}^{L-1} x(n)x(n+m-p) + \frac{1}{L\sigma_x\sigma_y}\sum_{n=0}^{K-1} x(n)n_2(n+m)$$
$$+\frac{1}{L\sigma_x\sigma_y}\sum_{p=0}^{K-1} h(p)\sum_{n=0}^{L-1} n_1(n)x(n+m-p) + \frac{1}{L\sigma_x\sigma_y}\sum_{n=0}^{K-1} n_1(n)n_2(n+m) \quad (2A.2)$$

The first term is the same as the one described in Chap. 2. The following three terms become negligible if there are no correlations among the input $x(n)$ and the noises $n_1(n)$ and $n_2(n)$. Therefore, the following equation holds.

$$r_{xy}(m) = \frac{L\sigma_x^2}{L\sigma_x\sigma_y} h(m) = \frac{\sigma_x}{\sigma_y} h(m) \qquad (2.11) \qquad (2A.3)$$

Appendix 2B Impulse Response Estimation in the Time Domain

It has been made clear in Chap. 2 that the cross-correlation between the input and the output becomes the impulse response of the system when a random sequence is input to the system. If the input and output sequences with a random input sequence are available, it is possible to estimate the transfer function by a simpler method, which will be explained here.

The output sequence $y(n)$ is given by the convolution of the input sequence $x(n)$ and the impulse response $h(n)$ as follows (see Chap. 1),

$$y(n) = \sum_{p=0}^{m-1} x(n-p)h(p) \qquad (1.1) \qquad (2B.1)$$

By dividing the both sides of the equation with nonzero term $x(n-r)$, we have

$$\frac{y(n)}{x(n-r)} = \sum_{p=0}^{r-1} \frac{x(n-p)}{x(n-r)} h(p) + h(r) + \sum_{p=r+1}^{m-1} \frac{x(n-p)}{x(n-r)} h(p) \qquad (2B.2)$$

By taking the averages of both sides of Eq. (2B.2), it is simplified as

$$\lim_{N\to\infty} \frac{1}{N}\sum_{n=1}^{N} \frac{y(n)}{x(n-r)} = h(r) \qquad (2B.3)$$

This is because $x(n-p)/x(n-r)$ is also a random sequence with zero mean.

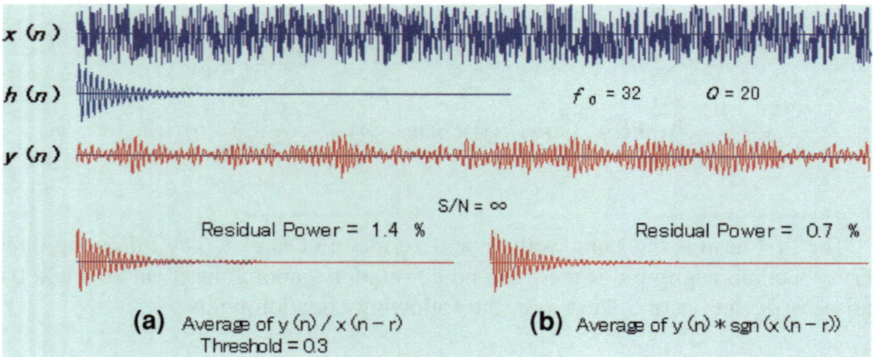

Fig. 2B.1 Example of impulse response estimation by using only the averaging operation. Animation available in supplementary files under filename E9-B1_IR-ESTIMATION.exe

If the mean of the input sequence is zero, the impulse response can be estimated using this equation. The only necessary requirement is that the mean of $x(n)$ is zero, and not that the amplitude of $x(n)$ must be random. Even if the input sequence has a constant amplitude with random signs, the impulse response can be estimated. Even for a random input sequence, only necessary information is the sign.

Therefore, for the case with the random input sequence or the constant-amplitude and random-sign sequence, the following equation without division can be used for the impulse response estimation.

$$\lim_{N\to\infty}\frac{1}{N}\sum_{n}^{N} y(n)\mathrm{sgn}\{x(n-r)\} = h(r) \tag{2B.4}$$

Figure 2B.1 shows an example of the impulse response estimation by the present method. The input sequence is random with maximum amplitude of 1. The input, impulse, and the output sequences are shown by the top three charts, and the impulse responses estimated by Eqs. (2B.3) and (2B.4) are shown by the bottom two charts. In both cases, the impulse response is estimated with high accuracy. In Fig. 2B.1a, which was obtained by using Eq. (2B.3), samples with amplitudes smaller than 0.3 (threshold) were removed in order to avoid divisions by small values.

You can change the input sequence, impulse response, and threshold in the program attached to the figure. It also tells you that this method cannot be applied to the cases with moving-averaged input sequences.

Appendix 2C Derivation of Eq. (2.15)

The autocorrelation function $R_{yy}(m)$ is calculated by

$$\begin{aligned}
R_{yy}(m) &= \frac{1}{L\sigma_y^2}\sum_{n=0}^{L-1} y(n)y(n+m) \\
&= \frac{1}{L\sigma_y^2}\sum_{n=0}^{L-1} \{x(n)+rx(n-d)\}\{x(n+m)+rx(n+m-d)\} \\
&= \frac{1}{L\sigma_y^2}\sum_{n=0}^{L-1} \{x(n)x(n+m)+rx(n-d)x(n+m)\} \\
&\qquad + \frac{1}{L\sigma_y^2}\sum_{n=0}^{L-1} \{rx(n)x(n+m-d)+r^2x(n-d)x(n+m-d)\} \\
&= R_{xx}(m)+rR_{xx}(m+d)+rR_{xx}(m-d)+r^2R_{xx}(m) \\
&= (1+r^2)R_{xx}(m)+rR_{xx}(m+d)+rR_{xx}(m-d) \qquad (2.15)
\end{aligned} \tag{2C.1}$$

Appendix 2D Residual Power Ratio of the Autocorrelation Functions Calculated by the FFT Method to the Autocorrelation Function Calculated by the Direct Method in the Time Domain

The (normalized) autocorrelation function of an infinitely long sine wave is a cosine wave with the amplitude 1 and with the same frequency as the original wave. The autocorrelation function of a sine wave with length T and 0 elsewhere is also a cosine wave with the same frequency, but its amplitude is 1 at τ(lag) = 0 and 0 at τ = T, changing linearly between them.

The autocorrelation function of the sine wave with length T mentioned above is obtained numerically by the 2N-point FFT/IFFT by adding N-point zero sequence to the N-point sine wave sequence. The result is the same as the one theoretically obtained.

Since the results of the N-point FFT/IFFT methods are used as the references, the range of interest is limited within $\pm N/2$. The residual power at time lag τ is given by

$$e(\tau) = |\tau|/T \tag{2D.1}$$

The integration of the residual power from $\tau = -N/2$ to $\tau = N/2$ gives the residual energy.

$$E = 2\int_0^{T/2} e^2(\tau)\,d\tau = \frac{2}{T^2}\frac{1}{3}\tau^3\Big|_0^{T/2} = \frac{T}{12} \tag{2D.2}$$

By normalizing the above equation by the energy $T \times 1 = T$, the residual energy (or average power) ratio becomes 1/12.

Appendix 2E Derivation of Eq. (2.27)

$$\begin{aligned} R_{xy}(k) &= \frac{1}{N\sigma_x\sigma_y}\sum_{p=0}^{N-1}\sum_{n=0}^{N-1} x^*(n)y(p)\exp(-j2\pi\frac{pk}{N})\exp(j2\pi\frac{nk}{N}) \\ &= \frac{1}{N\sigma_x\sigma_y}\sum_{n=0}^{N-1} x^*(n)\exp(j2\pi\frac{nk}{N})\sum_{p=0}^{N-1} y(p)\exp(-j2\pi\frac{pk}{N}) \end{aligned}$$

therefore,

$$R_{xy}(k) = \frac{1}{N\sigma_x\sigma_y}X^*(k)Y(k)(2.27) \tag{2E.1}$$

Appendix 2F Derivation of Eq. (2.31)

First, the Fourier transform of the cross-correlation is performed.

$$\begin{aligned} R_{xy}(f) &= \int_{-\infty}^{+\infty} r_{xy}(\tau)\exp(-j2\pi f\tau)d\tau \\ &= \frac{1}{\sigma_x\sigma_y}\int_{-\infty}^{+\infty}\int_{-\infty}^{+\infty} x^*(t)y(t+\tau)dt\exp(-j2\pi f\tau)d\tau \end{aligned}$$

Then, the order of integration is reversed.

$$\begin{aligned} R_{xy}(f) &= \frac{1}{\sigma_x\sigma_y}\int_{-\infty}^{+\infty} x^*(t)\int_{-\infty}^{+\infty} y(t+\tau)\exp(-j2\pi f\tau)d\tau dt \\ &= \frac{1}{\sigma_x\sigma_y}\int_{-\infty}^{+\infty} x^*(t)Y(f)\exp(j2\pi ft)dt \\ &= \frac{1}{\sigma_x\sigma_y}Y(f)\int_{-\infty}^{+\infty} x^*(t)\exp(j2\pi ft)dt \\ &= \frac{1}{\sigma_x\sigma_y}X^*(f)Y(f) \qquad (2.31) \end{aligned} \tag{2F.1}$$

Conversely, let us calculate the Inverse Fourier transform of the cross-spectrum.

$$\int_{-\infty}^{+\infty} X^*(f)Y(f)\exp(j2\pi ft)df = \int_{-\infty}^{+\infty} Y(f)\exp(j2\pi ft)\int_{-\infty}^{+\infty} x^*(t)\exp(-j2\pi ft)dtdf$$

By reversing the order of integration, we get:

$$\begin{aligned}\int_{-\infty}^{+\infty} X^*(f)Y(f)\exp(j2\pi ft)df &= \int_{-\infty}^{+\infty} x^*(t)\int_{-\infty}^{+\infty} Y(f)\exp\{j2\pi f(t+\tau)\}df\,dt \\ &= \int_{-\infty}^{+\infty} x^*(t)y(t+\tau)dt = r_{xy}(\tau)\end{aligned} \quad (2F.2)$$

Appendix 3B Reason for the Change of Zero(s) When One Sample in a Sample Sequence is Replaced by Another Value

Let the original sequence be $x_0(n)$ and its DFT be $X_0(k)$. It is assumed that $X_0(k_i) = 0$. Another sequence $x_1(n)$ is made by adding d ($\neq 0$) at $n = m$ ($0 \leq m \leq N-1$). The N-point DFT of a sequence with all zeros except for d at $n = m$ is given by $(d/N)e^{(-j2\pi mk/N)}$. Since the DFT $X_1(k)$ of $x_1(n)$ is given by the sum of $X_0(k)$ and $(d/N)e^{(-j2\pi mk/N)}$ because of the additional property, the value at $k = k_i$ is equal to $(d/N)e^{(-j2\pi mk/N)}$, which is of course not equal to zero, indicating that $X_1(k)$ does not have zero at $k = k_i$.

Appendix 4A Complex Cepstrum

The cepstrum discussed in Chap. 4 is the DFT of the logarithm of the power spectrum. Since the power spectrum is nonnegative value, only zeros of the power spectrum must be avoided in calculating the logarithm. However, the information on the phase is lost and only the rough shape of the transfer function can be estimated.

Instead of taking the logarithm of the power spectrum, the direct logarithm of the complex spectrum of a time waveform is the **complex cepstrum**.

We are trying to calculate the logarithm of Eq. (4.8). Since it is the product of the input spectrum and the system transfer function, its logarithm is the sum of the logarithms of each of them. However, each term of Eq. (4.8) is a complex number. “What is the logarithm of a complex number?” is a natural question. First, let us consider about this.

Eq. (4.8) is expressed by:

$$V(k) = V_R(k) + V_I(k) \quad (4A.1)$$

where $V_R(k)$ and $V_I(k)$ are the real and imaginary parts of $V(k)$, respectively. Then, the absolute value and argument are given, respectively, by

$$|V(k)| = \sqrt{V_R^2(k) + V_I^2(k)} \tag{4A.2}$$

and

$$\theta(k) = \arctan[V_I(k)/V_R(k)] \tag{4A.3}$$

Using these two equations, Eq. (4A.1) is expressed as

$$V(k) = |V(k)| \exp\{j\theta(k)\} \tag{4A.4}$$

Therefore, the natural logarithm of $V(k)$ becomes

$$\log\{V(k)\} = \log\{|V(k)|\} + j\theta(k) \tag{4A.5}$$

Now, it became possible to take the logarithm of $V(k)$, the DFT of the output $v(n)$.

If we express $X(k)$ and $H(k)$ by

$$X(k) = |X(k)| \exp\{j\phi(k)\}$$

and

$$H(k) = |H(k)| \exp\{j\beta(k)\}$$

we have

$$V(k) = X(k)H(k) = |X(k)||H(k)| \exp[j\{\phi(k) + \beta(k)\}] \tag{4A.6}$$

and

$$\log V(k) = \log|X(k)| + \log|H(k)| + j\{\phi(k) + \beta(k)\} \tag{4A.7}$$

The real and imaginary parts of Eq. (4A.7) is given, respectively, by

$$\log|V(k)| = \log|X(k)| + \log|H(k)| \tag{4A.8}$$

and

$$\theta(k) = \phi(k) + \beta(k) \tag{4A.9}$$

Equation (4A.8) is exactly the same as Eq. (4.10). In the actual calculation, the logarithm of the periodic power spectrum is used. Its DFT is the same as Eq. (4.11). As far as the real part is concerned, the complex cepstrum is the same as the cepstrum described in 4.3.

Equation (4A.9) shows the general rule that the phase of a product of two complex numbers are given by the sum of phases of the complex numbers. This also indicates that the phase response of the transfer function is obtained by

$$\beta(k) = \theta(k) - \phi(k) \tag{4A.10}$$

It is assumed in Chap. 4 that the phase of $x(n)$ is unobservable. In this case, $\phi(k)$ cannot be used to calculate $\beta(k)$.

Theoretically, if the impulse response of the system (IDFT of $H(k)$) is causal, the phase response of the transfer function is uniquely determined by the absolute value of the transfer function. However, this topic is not discussed in this book and it is not the author's intension to show the calculation method without reasoning. Also the shape of the transfer function obtained by the cepstrum method is too coarse. The cepstrum is theoretically interesting, but the author does not have successful experiences in practical applications.

Appendix 4B DFT of $x(n-d)$

$$\mathrm{DFT}[x(n-d)] = \sum_{n=0}^{N-1} x(n-d)\exp\left(-j2\pi\frac{nk}{N}\right)$$

By letting $p = n - d$ ($n = p + d$), and because of the circularity of the DFT, the summation from $n = 0$ to $N - 1$ can be replaced from $p = 0$ to $N - 1$. Then, the above equation becomes

$$\begin{aligned}\mathrm{DFT}[x(n-d)] &= \sum_{p=0}^{N-1} x(p)\exp\left\{-j2\pi\frac{(p+d)k}{N}\right\}\\ &= \sum_{p=0}^{N-1} x(p)\exp\left\{-j2\pi\frac{pk}{N}\right\}\exp\left\{-j2\pi\frac{dk}{N}\right\}\end{aligned}$$

Therefore, if the DFT of $x(n)$ is $X(k)$, the DFT of $x(n\text{-}d)$ is given by

$$\mathrm{DFT}[x(n-d)] = X(k)\exp\left\{-j2\pi\frac{dk}{N}\right\}$$

Appendix 5A Derivation of Eq. (5.15)

$$x(n) = \mathrm{IDFT}[X(k)] = \sum_{k=0}^{N-1} X(k)\exp(j2\pi\frac{kn}{N})$$

By letting, $X(k) = R(k) + jI(k)$

$$x(n) = \sum_{k=0}^{N-1} \{R(k) + jI(k)\} \left\{ \cos(2\pi \frac{kn}{N}) + j\sin(2\pi \frac{kn}{N}) \right\}$$
$$= \sum_{k=0}^{N-1} \left\{ R(k)\cos(2\pi \frac{kn}{N}) - I(k)\sin(2\pi \frac{kn}{N}) \right\}$$
$$+ j\sum_{k=0}^{N-1} \left\{ R(k)\sin(2\pi \frac{kn}{N}) + I(k)\cos(2\pi \frac{kn}{N}) \right\}$$

Since R(k) and cosine functions are even, and I(k) and sine functions are odd, the imaginary part becomes zero

$$x(n) = \sum_{k=0}^{N-1} \left\{ R(k)\cos(2\pi \frac{kn}{N}) - I(k)\sin(2\pi \frac{kn}{N}) \right\} \quad (5A.1)$$

Appendix 5B Check of Eq. (5.20)

Express the $x(n)$ and $x_{\perp}(n)$ by the summation from k = 0 to $N/2 - 1$ as

$$x(n) = \mathrm{IDFT}[X(k)] = \frac{1}{N} \sum_{k=-N/2}^{N/2-1} X(k)\exp(j2\pi \frac{kn}{N}) \quad (5.16)$$
$$= \frac{1}{N} \sum_{k=0}^{N/2-1} X(k)\exp(j2\pi \frac{kn}{N}) + \frac{1}{N} \sum_{k=0}^{N/2-1} X(-k)\exp(-j2\pi \frac{kn}{N})$$
$$= \frac{1}{N} \sum_{k=0}^{N/2-1} [R(k) + jI(k)]\exp(j2\pi \frac{kn}{N}) + \frac{1}{N} \sum_{k=0}^{N/2-1} [R(-k) + jI(-k)]\exp(-j2\pi \frac{kn}{N})$$
$$= \frac{1}{N} \sum_{k=0}^{N/2-1} [R(k) + jI(k)]\exp(j2\pi \frac{kn}{N}) + \frac{1}{N} \sum_{k=0}^{N/2-1} [R(k) - jI(k)]\exp(-j2\pi \frac{kn}{N})$$
$$= \frac{1}{N} \sum_{k=0}^{N/2-1} R(k)\left[\exp(j2\pi \frac{kn}{N}) + \exp(-j2\pi \frac{kn}{N})\right]$$
$$+ j\frac{1}{N} \sum_{k=0}^{N/2-1} I(k)\left[\exp(j2\pi \frac{kn}{N}) - \exp(-j2\pi \frac{kn}{N})\right]$$

Then $x(n)$ is given by

$$x(n) = \frac{2}{N} \sum_{k=0}^{N/2-1} R(k)\cos(2\pi \frac{kn}{N}) - \frac{2}{N} \sum_{k=0}^{N/2-1} I(k)\sin(2\pi \frac{kn}{N})$$

Similarly, $x_{\perp}(n)$ is given by

$$x_{\perp}(n) = \frac{2}{N}\sum_{k=0}^{N/2-1} I(k)\cos(2\pi\frac{kn}{N}) + \frac{2}{N}\sum_{k=0}^{N/2-1} R(k)\sin(2\pi\frac{kn}{N})$$

In the summation of products of these two equations, sums of products of cos and sin terms are zero, and sums of products of same cos terms and sums of products of the same sin terms are zero if ks are different. Therefore, only $\cos^2$ and $\sin^2$ terms with the same k remain as

$$\begin{aligned}\sum_{n=0}^{N-1} x(n)x_{\perp}(n) &= \frac{4}{N^2}\sum_{n=0}^{N-1}\left[\sum_{k=0}^{N/2-1} R(k)I(k)\cos^2(2\pi\frac{kn}{N}) - \sum_{k=0}^{N/2-1} R(k)I(k)\sin^2(2\pi\frac{kn}{N})\right]\\ &= \frac{4}{N^2}\left[\sum_{k=0}^{N/2-1} R(k)I(k)\sum_{n=0}^{N-1}\cos^2(2\pi\frac{kn}{N}) - \sum_{k=0}^{N/2-1} R(k)I(k)\sum_{n=0}^{N-1}\sin^2(2\pi\frac{kn}{N})\right]\\ &= \frac{2}{N^2}\left[\sum_{k=0}^{N/2-1} R(k)I(k)\{N-N\}\right] = 0 \qquad (5.20)\end{aligned}$$

(5B.1)

because

$$\cos^2(2\pi\frac{kn}{N}) = \frac{1}{2}\left\{1+\cos(4\pi\frac{kn}{N})\right\}$$

$$\sin^2(2\pi\frac{kn}{N}) = \frac{1}{2}\left\{1-\cos(4\pi\frac{kn}{N})\right\}$$

References

1. N. Aoshima, Computer-generated pulse applied for sound measurement. J. Acoust. Soc. Am. **69**(5), 1484–1488 (1981)
2. Y. Suzuki, F. Asano, H-Y. Kim, T. Sone, An optimum computer- generated pulse signal suitable for the measurement of very long impulse responses. J. Acoust. Soc. Am. **97**(2), 1119–1123 (1995)
3. B.P. Bogert, M.J.R. Healy, J.W. Tukey, in *The Quefrency Analysis of Time Series for Echoes: Cepstrum, Pseudo-autocovariance, Cross Cepstrum, and Saphe Cracking*, ed. by M. Rosenblatt, Proceedings on Symposium Time Series Analysis (Wiley, New York, 1963), pp. 209–243
4. A. Papoulis, *Signal Analysis* (McGrawhill. Book Co., New York, 1977)
5. N.A. Jacobson, *Basic Algebra I* (W. H. Freeman and Company, New York, 1985)
6. C.E. Shannon, The mathematical theory of communication. Bell Syst. Tech. J. **27**(379-423), 623–656 (1948)
7. N. Wiener, *Extrapolation, Interpolation and Smoothing of Stationary time Series with Engineering Application* (MIT Press, Cambridge, MA, 1949)
8. A.V. Oppemheim, R.W. Schafer, *Digital Signal Processing*. (Prentice Hall, Englewood Cliffs, New Jersey, 1975)
9. L.R. Rabiner, B. Gold, *Theory and Application of Digital Signal Processing*. (Prentice Hall, Englewood Cliffs, New Jersey, 1975)
10. J.D. Markel, A.H. Gray Jr., *Linear Prediction of Speech*. (Springer, New York, 1976)

K. Kido, *Digital Fourier Analysis: Advanced Techniques*,
DOI: 10.1007/978-1-4939-1127-1,

Answers

Chapter 1

1.

$$y(n) = \sum_{p=n-m+1}^{n} x(p)h(n-p) \tag{1.2}$$

Replace the variable by $q = n - p$ $(p = n - q)$. Since $q = m - 1$ for $p = n - m + 1$ and $q = 0$ for $p = n$, Eq. (1.17) is given by

$$y(n) = \sum_{q=m-1}^{0} x(n-q)h(q)$$

By changing the order of summation, Eq. (1.1) is obtained.

2. *NM*
3. *M*
4. If $p \neq m$, the product is 0, and if $p = m$, there is only one term for discrete frequency p (=m).
5. Even if $p \neq m$, the product is not zero, and if $p = m$, there are other frequency components as well as at p.
6. If $M < N$, divide x(n) into M-point sequences and add M-point zeros to each of them and then calculate the convolution by the 2M-point FFTs. If $N < M$, divide h(n) into N-point sequences and calculate the convolution by the 2M-point FFTs.
7. Choose the first 7N-point data from x(n) and add N-point zeros to make an 8N-point sequence. Add 7N-point zeros to h(n) to make an 8N-point sequence. Use 8N-point FFT and obtain 8N-point output. Choose the next 7N-point data from x(n) and calculate the output the same way. Add the last N-point data from the previous output to the present first N-point output data.

K. Kido, *Digital Fourier Analysis: Advanced Techniques*,
DOI: 10.1007/978-1-4939-1127-1,

Chapter 2

1.

$$1/\sqrt{2}$$

2. Because the mean of x(n)y(n) is zero.
3. Because the second term in the parenthesis is -1.
4. Substitute Eq. (1.6) into Eq. (2.22).

$$r_{xy}(\tau) = \frac{1}{\sigma_x \sigma_y} \int_{-\infty}^{+\infty} x^*(t) \int_{-\infty}^{+\infty} x(t+\tau-m)h(m)dmdt$$

Change the order of integration.
Since, if x(t) is a random signal with zero mean, the integration with respect to t is σ_x^2 for $\tau = m$ and 0 for $\tau \neq m$,

$$r_{xy}(\tau) = \frac{1}{\sigma_x \sigma_y} \int_{-\infty}^{+\infty} h(m)\sigma_x^2 dm = \frac{\sigma_x}{\sigma_y} h(\tau)$$

5. Yes.
6.

$$T_{1/2} = \left(\frac{\log 2}{\log e}\right) T = 0.693$$

7. 1 for $\tau = 0$ and 0 for $\tau = \pm T(\tau : \text{lag})$.
8.

$$\cos(2\pi\tau/T)$$

9. 0
10.

$$\frac{1}{3\pi} \cos 3\pi \frac{\tau}{T}$$

11.

$$-\sin(2\pi\tau/T)$$

12.

$$-\sin(3\pi\tau/T)$$

13.

$$\frac{1}{7.6\pi}[1-\cos(7.6\pi)]\cos(2\pi\frac{2.3\tau}{T})+\frac{1}{7.6\pi}\sin(7.6\pi)\sin(2\pi\frac{2.3\tau}{T})$$
$$-\frac{1}{1.6\pi}[1-\cos(1.6\pi)]\cos(2\pi\frac{2.3\tau}{T})-\frac{1}{1.6\pi}\sin(1.6\pi)\sin(2\pi\frac{2.3\tau}{T})$$

14. (1) When the signal to be analyzed contains a periodic component and P equals an integer multiple of the period or equals the inverse of the integer multiple of the period, or (2) when an external noise contains a periodic component and P equals an integer multiple of the period or equals the inverse of the integer multiple of the period.

Chapter 3

1. X(f) is complex such that $X(f) = X_R(f) + jX_I(f)$. Its complex conjugate is $X*(f) = X_R(f) - jX_I(f)$.
   Then,$X*(f)X(f) = \{X_R(f) - jX_I(f)\}\{X_R(f) + jX_I(f)\} = X_R^2(f) + X_I^2(f)$.
2. It is clear that $W_{XX}(f)$ is real from the answer 1.

$$X*(f)Y(f) = \{X_R(f) - jX_I(f)\}\{Y_R(f) + jY_I(f)\} = X_R(f)Y_R(f) + X_I(f)Y_I(f) + j\{X_R(f)Y_I(f) - X_I(f)Y_R(f)\}$$

   This is a complex number except for the special case with $X_R(f)Y_I(f) - X_I(f)Y_R(f)$.
3. Fourier transform pair.
4. Fourier transform pair.
5. x(t) and n(t) are independent.
6. It is a function that defines the relation between the input and the output of a transfer system in the frequency domain. It is the Fourier transform of an impulse response, and it is given by the ratio of the output spectrum to the input spectrum.
7. It is a noise-free N-point output that circulates with period N.
8. Add a zero sequence, which is not shorter than the impulse response, to the input sequence, and assume that it is the N-point input sequence. Also, add another zero sequence to the impulse response so that the total length equals N and use its DFT as the transfer function.
9. Estimated transfer function. It has a larger error when the impulse response is longer, in other words, when it has a slower decay (see Fig. 3.3).
10. A time-reversed impulse response appears at the end of the estimated impulse response. The amplitude of the time-reversed impulse response is proportional to the degree of the correlation.
11. Necessary (check with the program attached to Fig. 3.4).

12. (a) The random input does not have a correlation.
    (b) The FFT size, N, must be larger than or equal to the impulse response length, L.
    (c) To use an input sequence made of the first N-L point data and L-point zeros, and to use the whole N-point output.
    (d) A large umber of averaging.
13. Necessary in order to reduce the effect of external noise. If there is noise, the averaging is not necessary.
14. Refer to the text.

Chapter 4

1. Calculate the cross-correlation function with a signal near the source, or calculate the impulse response from the IDFT of the transfer function obtained by the cross-spectrum method.
2. $$\mathrm{DFT}[x(n-d)] = X(k)\exp\{-j2\pi\frac{dk}{N}\}$$
3. 32 = 1024/32
4. When the amplitude of the fundamental component is smaller than those of other components.
5. (5), (2), (4)
6. When the external random noise is large.
7. A delay of a signal causes a phase delay which is proportional to the multiple of the delay time and the frequency. This causes the periodicity in the spectrum. To be able to observe the periodicity, the spectrum should be widely spread.
8. Refer to 4.2.
9. Refer to 4.3.
10. The signal should have frequency components in the wide frequency range.

Chapter 5

1. 0
2. The same way as to derive Eq. (2.24) from Eq. (2.21).
3. The same way as to derive Eq. (2.24) from Eq. (2.21).
4. 4. $x_{\perp}(t) = -\cos(2\pi f_0 t)$
5. A signal made of two cosine waves with frequencies ω and $b\omega$ is given by
   $$x(t) = \cos(\omega t) + A\cos(b\omega t)$$

An orthogonal wave to this signal is given by

$$x_{\perp}(t) = \sin(\omega t) + A\sin(b\omega t)$$

The instantaneous phase of a signal with the real part $x(t)$ and the imaginary part $x_{\perp}(t)$ is given by

$$\phi(t) = \tan^{-1}\frac{x_{\perp}(t)}{x(t)} = \tan^{-1}\frac{\sin(\omega t) + A\sin(b\omega t)}{\cos(\omega t) + A\cos(b\omega t)}$$

The instantaneous frequency is obtained by the time derivative of the phase.

$$f(t) = \frac{1}{2\pi}\frac{d}{dt}\phi(t) = \frac{1}{2\pi}\frac{d}{dt}(\tan^{-1} y)$$

where

$$y = \frac{\sin(\omega t) + A\sin(b\omega t)}{\cos(\omega t) + A\cos(b\omega t)}$$

Then

$$f(t) = \frac{1}{2\pi}\frac{d\phi(t)}{dt} = \frac{1}{2\pi}\frac{d}{dy}\tan^{-1}(y)\frac{dy}{dt}$$

Each derivative above becomes

$$\frac{dy}{dt} = \frac{d}{dt}\frac{\sin(\omega t) + A\sin(b\omega t)}{\cos(\omega t) + A\cos(b\omega t)} = \omega\frac{1 + A^2 b + A(1+b)\cos\{(1-b)\omega t\}}{\{\cos(\omega t) + A\cos(b\omega t)\}^2}$$

and

$$\frac{d}{dy}\tan^{-1} y = \frac{1}{1+y^2} = \frac{\{\cos(\omega t) + A\cos(b\omega t)\}^2}{1 + A^2 + 2A\cos\{(1-b)\omega t\}}$$

After several steps, we have

$$f(t) = \frac{\omega}{2\pi}\frac{1 + A^2 b + A(1+b)\cos\{(1-b)\omega t\}}{1 + A^2 + 2A\cos\{(1-b)\omega t\}}$$

Since the denominator is positive (except for very special cases), the instantaneous frequency takes negative values between two values of A which satisfy

$$1 + A^2 b - A(1+b) = 0$$

or

$$A = \frac{(1+b) \pm |1-b|}{2b}$$

For example, the instantaneous frequency becomes negative within (1/2, 1) for b = 2, and within (1/3, 1) for b = 3. You can check these by running the program attached to Fig. 5.9.

6. Included in answer 5.
7. Since the instantaneous phase of the sum of the two sidebands coincide with the instantaneous phase of the carrier, the amplitude-modulated signal has a constant instantaneous frequency.
8. Let $\tilde{x}(t)$ be a function that has the same positive frequency spectra as $x(t)$ and zero negative frequency spectra. Then it can be expressed as:

$$\begin{aligned}\tilde{x}(t) &= \frac{1}{2}\mathrm{IFT}[X(f) + \mathrm{sgn}(f)X(f)] \\ &= \frac{1}{2}\int_{-\infty}^{+\infty} X(f)\exp(j2\pi ft)df + \frac{1}{2}\int_{-\infty}^{+\infty} \mathrm{sgn}(f)X(f)\exp(j2\pi ft)df\end{aligned}$$

Since

$$\mathrm{IFT}[\mathrm{sgn}(f)] = j\frac{1}{\pi t}$$

$\tilde{x}(t)$is given by

$$\begin{aligned}\tilde{x}(t) &= \frac{1}{2}x(t) + j\frac{1}{2\pi}\int_{-\infty}^{+\infty} \frac{1}{t-\tau}x(\tau)d\tau \\ \tilde{x}(t) &= \frac{1}{2}[x(t) + jx_{\perp}(t)]\end{aligned}$$

where $x_{\perp}(t)$ is the Hilbert transform of $x(t)$.

9. The complex signal has zero negative spectra. The reason is given in answer 8.

Chapter 6

1. For every μ and v, $X(v,\mu) = 1$.
2. For every μ, $X(0,\mu) = 1$. For nonzero v, $X(v,\mu) = 0$. You can check these two problems by using Fig. 6.6.
3. Same as above.
4. $X(0,0) = N^2 + 1. X(v,\mu) = 1$ for all other μ and v.

5.

$$Z(\nu, m) = \sum_{n=0}^{N-1} x(n, m) \exp\left(-j2\pi \frac{\nu \cdot n}{N}\right) \tag{6A.1}$$

$$X(\nu, m) = \sum_{m=0}^{M-1} Z(\nu, m) \exp\left(-j2\pi \frac{\mu \cdot m}{M}\right) \tag{6A.2}$$

$$X(\nu, m) = \sum_{n=0}^{N-1} \sum_{m=0}^{M-1} x(n, m) \exp\left(-j2\pi \frac{\mu \cdot m}{M}\right) \exp\left(-j2\pi \frac{\nu \cdot n}{N}\right)$$

6. Same as deriving Eqs. (6.3) and (6.4).
7. Obtain the answer using the program attached to Fig. 6.14.
8. Same as answer 7.
9. When you can divide the original image into smaller sub-images and if the range of the density variation of each subimage is smaller than that of the original image.
10. The DCTs handle only real value.

Index

K. Kido, *Digital Fourier Analysis: Advanced Techniques*,
DOI: 10.1007/978-1-4939-1127-1,

GPSR Compliance

The European Union's (EU) General Product Safety Regulation (GPSR) is a set of rules that requires consumer products to be safe and our obligations to ensure this.

If you have any concerns about our products, you can contact us on ProductSafety@springernature.com

In case Publisher is established outside the EU, the EU authorized representative is:

Springer Nature Customer Service Center GmbH
Europaplatz 3
69115 Heidelberg, Germany

Batch number: 10607224

Printed by Printforce, the Netherlands